Transaction-Level Power Modeling

Amr Baher Darwish · Magdy Ali El-Moursy ·
Mohamed Amin Dessouky

Transaction-Level Power Modeling

 Springer

Amr Baher Darwish
Mentor a Siemens Business
Cairo, Egypt

Magdy Ali El-Moursy
Mentor a Siemens Business
Cairo, Egypt

Mohamed Amin Dessouky
Ain Shams University
Cairo, Egypt

ISBN 978-3-030-24829-1   ISBN 978-3-030-24827-7   (eBook)
https://doi.org/10.1007/978-3-030-24827-7

This Springer imprint is published by the registered company Springer Nature Switzerland AG
The registered company address is: Gewerbestrasse 11, 6330 Cham, Switzerland

Preface

Transaction-Level Modeling (TLM) is a technique for building and developing designs. TLM introduces abstract modeling of communication schemes between design modules. It provides insights for system-level design at early stages of development. Thus, TLM accelerates the design cycle of System-on-Chip (SoC) in the industry.

A novel methodology, referred to as Transaction-Level Power Modeling (TLPM), is proposed. It is used to evaluate power consumption using TLM. Evaluating power consumption at early phases of product design cycle is important to decrease the number of expensive design iterations. The methodology exploits the existing tools for Register Transfer Level (RTL) simulation, design synthesis, and SystemC prototyping to provide fast and accurate power results. TLPM is performed in two stages: power characterization on RTL and power modeling on TLM. Experimental results reveal both efficiency and accuracy of the TLPM methodology. TLPM speeds up the simulation time by more than two orders of magnitude over RTL, while the error in power estimation is less than 3%.

Cairo, Egypt

Amr Baher Darwish
Magdy Ali El-Moursy
Mohamed Amin Dessouky

Acknowledgements

I would like to express my great gratitude to Prof. Mohamed Dessouky and Dr. Magdy El-Moursy for their continuous guidance and insightful thoughts.

Special thanks to Karim Abo El-Makarem (may his soul rest in peace), Ahmed Aly, AbdelRahman Anwar, Adham Rageh, Ahmed El-Zeiny, and Ahmed Khalil for their contribution and qualification of methodology on different designs.

Finally, I am grateful to my family for assistance and inspiration to accomplish this book.

Contents

1	Introduction	1
	1.1 Motivation	1
	1.2 Contribution	2
	1.2.1 Power Characterization	2
	1.2.2 Power Modeling in TLM	3
	1.2.3 TLPM Methodology	3
	1.3 Book Summary	3
	References	4

Part I Background

2	Fundamental Concepts	9
	2.1 Introduction	9
	2.2 Design Modeling	9
	2.2.1 Introduction	9
	2.2.2 Design Abstraction Levels	10
	2.2.3 TLM Design Abstraction	11
	2.3 SystemC	12
	2.3.1 Scope	12
	2.3.2 Purpose	12
	2.3.3 SystemC Example	12
	2.4 Transaction-Level Modeling	14
	2.4.1 Overview	14
	2.4.2 TLM 2.0 Interfaces	26
	2.4.3 TLM-2.0 Global Quantum	34
	2.4.4 Combined TLM-2.0 Interfaces	34
	2.4.5 TLM-2.0 Sockets	34
	2.4.6 TLM-2.0 Phases	35
	2.4.7 Base Protocol	35
	2.4.8 TLM-2.0 Examples	36

2.5   Universal Verification Methodology.   39
      2.5.1   Introduction.   39
      2.5.2   Base Classes   40
      2.5.3   UVM Factory   41
      2.5.4   Predefined Classes   42
2.6   Design Synthesis   43
2.7   Switching Activity Interchange Format   44
2.8   Summary   44
References   45

3   Power Modeling and Characterization   47
3.1   Introduction   47
3.2   Power Modeling at Different Levels   47
      3.2.1   System Level   47
      3.2.2   Component Level   48
3.3   Levels of Abstraction in Modeling.   49
3.4   Power Characterization   50
      3.4.1   Gate-Level Netlist Characterization   50
      3.4.2   Hardware Characterization   50
      3.4.3   Datasheet-Based Characterization   51
      3.4.4   Look-up Tables' Characterization   52
3.5   Performance Overhead   52
3.6   Power Evaluation   52
      3.6.1   Power Analysis Flow.   53
3.7   Summary   54
References   55

Part II   Transaction-Level Power Modeling

4   Transaction-Level Power Modeling Methodology   61
4.1   Introduction   61
4.2   TLPM Flow   61
      4.2.1   Overview   61
      4.2.2   Power Estimation Using TLPM   62
4.3   Power Characterization   63
      4.3.1   TLPM Power Characterization Flow   63
      4.3.2   Correlation Matrix Example   66
      4.3.3   EDA Tools for Power Characterization   66
4.4   TLPM Implementation and Simulation.   72
      4.4.1   TLPM Implementation and Simulation Flow   73
      4.4.2   Power Model Example.   80
      4.4.3   EDA Tools for TLPM Implementation Phase   81
4.5   Summary   85
References   86

5   Experimental Results   87
   5.1   Introduction   87
   5.2   Design Environment   87
   5.3   Timer SP804 Experiment   87
      5.3.1   Design Information   87
      5.3.2   Design Configuration   88
      5.3.3   Design Interface   90
      5.3.4   Design Signals   90
      5.3.5   Registers of the Design   91
      5.3.6   TLPM Results   92
   5.4   ZYNQ-7000 SoC Experiment   96
      5.4.1   Design Information   96
      5.4.2   TLPM Results 102
   5.5   Summary 106
   References 106

6   Conclusions and Future Work 107
   6.1   Conclusions 107
   6.2   Future Work 108

Index 109

About the Authors

Amr Baher Darwish received his B.Sc. and M.Sc. degrees in Electronics Engineering from Ain Shams University, Cairo, Egypt, in 2011 and 2017, respectively. In 2011, he worked as Design Application Engineer for RF/AMS team, Intel Corporation, Cairo, Egypt. Between November 2013 and May 2016, he was Quality Assurance Engineer in IC Verification Solutions Department at Mentor, a Siemens business, Cairo, Egypt. During June 2016 and April 2018, he worked as Backline Customer Support Engineer in the same company. Between October 2017 and March 2019, he worked as Senior Quality Assurance Engineer in Mentor. Currently, he is Questa SIM Product Engineer at Mentor, a Siemens business, Cairo, Egypt. He has published journal and conference papers in dynamic power estimation.

Magdy Ali El-Moursy received the B.S. degree in electronics and communications engineering (with honors) and the Master's degree in computer networks from Cairo University, Cairo, Egypt, in 1996 and 2000, respectively, and the Master's and the Ph.D. degrees in electrical engineering in the area of high-performance VLSI/IC design from University of Rochester, Rochester, NY, USA, in 2002 and 2004, respectively. In summer of 2003, he was with STMicroelectronics, Advanced System Technology, San Diego, CA, USA. Between September 2004 and September 2006 he was a Senior Design Engineer at Portland Technology Development, Intel Corporation, Hillsboro, OR, USA. During September 2006 and February 2008 he was assistant professor in the Information Engineering and Technology Department of the German University in Cairo (GUC), Cairo, Egypt. Between February 2008 and October 2010 he was Technical Lead in the Mentor Hardware Emulation Division, Mentor Graphics Corporation, Cairo, Egypt.

Dr. El-Moursy is currently Engineering Manager in Integrated Circuits Verification Systems Division, Mentor, A Siemens Business and Associate Professor in the Microelectronics Department, Electronics Research Institute, Cairo, Egypt. He is Associate Editor in the Editorial Board of Elsevier Microelectronics Journal, Journal of Circuits, Systems, and Computers, and International Journal of Circuits and Architecture Design and Technical Program Committee of many IEEE

Conferences such as ISCAS, ICAINA, PacRim CCCSP, ISESD, SIECPC, and IDT. His research interest is in Networks-on-Chip/System-on-Chip, interconnect design and related circuit level issues in high performance VLSI circuits, clock distribution network design, digital ASIC circuit design, VLSI/SoC/NoC design and validation/verification, circuit verification and testing and low power design. He is the author of around 90 papers, six book chapters, and four books in the fields of high speed and low power CMOS design techniques and NoC/SoC.

Mohamed Amin Dessouky received the B.Sc. and M.Sc. degrees in electrical engineering from the University of Ain Shams, Cairo, Egypt, in 1992 and 1995, respectively, and the Ph.D. degree in electrical engineering from the University of Paris VI, Paris, France, in 2001. In 1992 he joined the Electronics and Electrical Communications Engineering Department, University of Ain Shams, where he is now a full Professor. From 2010 to 2013 he was the director of the Integrated Circuits Lab at the same department. Prof. Dessouky was a visiting professor at the University of Paris VI in 2002 and 2004. From 2004 to 2010, he was on leave to Mentor Graphics Egypt, where he has been leading a mixed-signal design team responsible for the design of high-speed serial links. He also participated in the research and development of an EDA tool for technology porting of analog circuit designs. He is currently a staff Engineer at the same company. He also served in the Technical Committees of many IEEE Conferences such as DATE, SMACD, IDT and ICM. His research interests include custom digital design, ultra-low voltage design, analog-to-digital converters and CAD for analog and mixed-signal design. Prof. Dessouky holds four US patents and has published several journal and conference papers in addition to a book chapter on analog layout design porting.

Abbreviations

AMBA	Advanced Microcontroller Bus Architecture
AP SoC	All Programmable SoC
APB	Advanced Peripheral Bus
AT	Approximately-time
CA	Cycle-Accurate
DMI	Direct Memory Interface
DUT	Design Under Test
EB	Energy per Byte
EMIO	Extended Multiplexed I/O
ESL	Electronic System Level
FPGA	Field-Programmable Gate Array
FRC	Free Running Counter
GL	Gate Level
GPIO	General Purpose I/O
GUI	Graphical User Interface
HDL	Hardware Description Language
HTLP	Hierarchical Transaction Level Power
HW	Hardware
I2C	Inter Integrated Circuit
IC	Integrated Circuit
LT	Loosely-timed
MIO	Multiplexed I/O
MISO	Master Input Slave Output
MOSI	Master Output Slave Input
OOP	Object-Oriented Programming
PKtool	Power Kernel Tool
PL	Programmable Logic
PPC	Performance Optimization With Enhanced RISC – Performance Computing
PS	Processing System

PV	Programmer View
PVT	Programmer View Plus Timing
RTL	Register Transfer Level
RXFIFO	Receiver First In First Out
SAIF	Switching Activity Interchange Format
SoC	System-on-Chip
SPI	Serial Peripheral Interface
SW	Software
TLM	Transaction Level Modeling
TLPM	Transaction Level Power Modeling
TXFIFO	Transmitter First In First Out
UART	Universal Asynchronous Receiver Transmitter
UPF	Unified Power Format
UVM	Universal Verification Methodology
VCD	Value Change Dump

List of Figures

Fig. 2.1 Number of transistors in Intel microprocessors [3] 10
Fig. 2.2 Different levels of design modeling [4]. 11
Fig. 2.3 Efficiency of modeling strategies [8]. 11
Fig. 2.4 System-level design cycle [7] 13
Fig. 2.5 SystemC example 13
Fig. 2.6 TLM classes [7] 15
Fig. 2.7 TLM use cases [7] 16
Fig. 2.8 TLM components [5] 21
Fig. 2.9 TLM initiator, interconnect, and target [5] 22
Fig. 2.10 TLM levels of abstraction [8]. 25
Fig. 2.11 Blocking transport [5]. 27
Fig. 2.12 Temporal decoupling [5]. 28
Fig. 2.13 Time quantum [5]. 29
Fig. 2.14 Message sequence—backward path [5]. 30
Fig. 2.15 Message sequence—return path [5] 30
Fig. 2.16 Message sequence—early completion [5] 31
Fig. 2.17 Message sequence—timing annotation [5] 32
Fig. 2.18 Transition sequence example [5]. 37
Fig. 2.19 Blocking interface method causality [5] 38
Fig. 2.20 Non-blocking interface method causality [5]. 39
Fig. 2.21 UVM flow [15]. 40
Fig. 2.22 SAIF file syntax format 44
Fig. 3.1 HTLP power estimation flow [1] 48
Fig. 3.2 Power kernel tool [16] 49
Fig. 3.3 HTLP power characterization flow [1] 51
Fig. 3.4 Power analysis flowchart 54
Fig. 4.1 Transaction-level power modeling flow 62
Fig. 4.2 Power estimation for large SoCs using TLPM 63
Fig. 4.3 TLPM power characterization process and EDA tools 64
Fig. 4.4 Characterization process 64
Fig. 4.5 Perl script snippet for SAIF file processing. 67

Fig. 4.6   Correlation example—RTL design   67
Fig. 4.7   QuestaSim flow   69
Fig. 4.8   Design compiler flow   70
Fig. 4.9   Power report.   71
Fig. 4.10   TLPM implementation process   72
Fig. 4.11   Power model—tracking of register changes   75
Fig. 4.12   Power model—evaluation of power   76
Fig. 4.13   Debug port example   78
Fig. 4.14   Stand-alone function for non-TLM signals   79
Fig. 4.15   Power model.   80
Fig. 4.16   Visualizer drivers/receivers window   82
Fig. 4.17   Visualizer Logic Cone window.   83
Fig. 4.18   Visualizer Time Cone window   83
Fig. 4.19   Vista flow.   84
Fig. 5.1   ARM Dual-Timer Module (SP804).   88
Fig. 5.2   Schematic for ARM Dual-Timer Module (SP804)   89
Fig. 5.3   AMBA APB signals.   91
Fig. 5.4   Dynamic power estimation for different operating
            conditions.   94
Fig. 5.5   Dynamic power estimation absolute error between
            TLPM and RTL   94
Fig. 5.6   Simulation time for different operating conditions   95
Fig. 5.7   Simulation time comparison between TLPM and RTL.   95
Fig. 5.8   ZYNQ-7000 platform   96
Fig. 5.9   GPIO block diagram.   97
Fig. 5.10   GPIO channel   98
Fig. 5.11   SPI block diagram   99
Fig. 5.12   Baud-rate generator of I2C   100
Fig. 5.13   I2C block diagram   100
Fig. 5.14   UART block diagram   101
Fig. 5.15   Baud-rate generator of UART.   102

List of Tables

Table 2.1   Permitted phase transitions [5].   36

Table 4.1   Correlation matrix example   68

Table 5.1   UMC technology library   88

Table 5.2   Test plan for ARM Timer SP804.   92

Table 5.3   Test scenarios   93

Table 5.4   Dynamic power estimation for different operation conditions
of GPIO.   104

Table 5.5   Dynamic power estimation for different operation conditions
of SPI   104

Table 5.6   Dynamic power estimation for different operation conditions
of I2C   104

Table 5.7   Dynamic power estimation for different operation conditions
of UART.   105

Table 5.8   Dynamic power estimation for different operation conditions
of ZYNQ-7000   105

Table 5.9   Simulation time for different operating scenarios
of ZYNQ-7000   106

List of Algorithms

Algorithm 4.1   Conditional Correlation Example   65
Algorithm 4.2   Power Model Example   81

Chapter 1
Introduction

1.1 Motivation

The advance in processing technologies has driven the complexity of electronic design. The shrinkage of the transistor size makes the design models sophisticated, thus the simulation time of those models is prolonged. This advance in technology is affecting dramatically the product design cycle.

Meanwhile, low power consumption is a fundamental target in Integrated Circuits (IC) industry. Huge data centers are characterized by operation of billions of gates. Simple power increase in each gate boosts the overall power consumption and hence the cost tremendously. At the same time, battery-based portable devices require efficient handling of power consumption [1, 2].

Designs are conventionally implemented on several levels of abstraction such as Register Transfer Level (RTL) or Gate Level (GL). Simulations on these levels are very time-consuming for large System-on-Chip (SoC) designs. Thus, determining the power of such large designs is very computational and expensive on both GL and RTL, if even possible. Power consumption should be addressed in early phases of the design process to prevent long expensive iterations.

Different methodologies and modeling techniques have been developed to keep the design cycle short through different variants [3, 4]. Transaction-Level Modeling (TLM) is one of those modeling techniques [5, 6]. SystemC [7] is a common Hardware Description Language (HDL) used for TLM. TLM introduces abstract modeling of the communication scheme between design modules using function calls [8]. Abstracted modeling using function calls makes TLM [9] approach an important candidate for speeding up the design process.

Integration of power computation with TLM is gaining interest in both academia and industry. TLM is needed for power estimation to detect any unexpected excessive power dissipation in the design at early stages. In this book, TLM is adopted for estimating power consumption in IC design. The approach is based on the existing commercial tools for RTL simulation, design synthesis and SystemC prototyping to build a framework for power modeling at transaction abstraction level.

© Springer Nature Switzerland AG 2020

A. B. Darwish et al., *Transaction-Level Power Modeling*,
https://doi.org/10.1007/978-3-030-24827-7_1

Achieving low-power design can be addressed at different levels of design process: transistor, logic circuit, architecture, or system [10, 11]. For transistor level, proper scaling for transistor size and operating voltage is needed. This scaling achieves better threshold operating voltage and lower parasitics. For low-power design on circuit level, bus and logic optimizations are performed. Those optimizations yield performance improvements during simulation. Power management of the circuit components is performed at the architectural level. This is achieved by classifying those components into power modules and shutting down supply voltage for them when not in use. On system level, the engineers can minimize the power at early stages of design [12]. This is performed using suitable high-level integration for design components according to their estimated power numbers.

The typical front-end design flow in the industry starts with a behavioral model of the design and ends with the corresponding gate level netlist [13]. The fully synthesized design is used to verify the exact functionality pre-fabrication against design specification. The netlist contains all gate parasitics and needed timing information. However, GL is inconvenient in addressing design issues in early stages, because it is very time-consuming to simulate [14]. On the other hand, RTL has implementation details for the design and tracks bus cycles precisely without the need for a full GL netlist [15]. Thus, RTL could be more suitable to simulate and verify modifications of small designs. But, it is not the best candidate in case of verification of large SoCs using several iterations.

In case of large SoCs, simulation is needed on a higher level of abstraction. TLM represents a behavioral model with minimal implementation details [16]. TLM is characterized by fast simulation time, making it a good candidate for SoC design at early stages [17]. TLM is a highly abstracted model for a design; the design components are modeled at their functional level. TLM is a suitable prototype for early design validation [18], as it allows initial reliable design exploration [19]. Thus, power analysis and estimation at TLM level are desirable.

1.2   Contribution

The contribution of the proposed Transaction-Level Power Modeling (TLPM) can be resumed in the following:

1.2.1   Power Characterization

Power characterization is the extraction of power parameters from a design. This encloses the exerted energy by different design components. This process adopts an existing commercial flow for power evaluation on RTL which emphasizes the feasibility. The final objective is to create a database of design elements with the corresponding energy numbers for different design settings. This is performed under the assumption that RTL design is already available.

1.2.2 Power Modeling in TLM

Power evaluation in this book is based on TLM ports. Power models are created for every read/write operation of each transaction. Monitors are added to the callback functions of design registers to track the activity of different elements. Hence, exerted energy is tracked for every operation, and the whole energy profile is calculated.

1.2.3 TLPM Methodology

The book provides a Transaction-Level Power Modeling (TLPM) methodology. The methodology integrates the power characterization process with the TLM design containing the power models. It is used for dynamic power estimation of SoC. The methodology is applied at the component level, in order to estimate power information for those building blocks and their interconnects. The whole flow is described in details in the following chapters including experimental results.

1.3 Book Summary

This book demonstrates power computation challenges for SoC design. It presents a full overview of the existing flows and methodologies for addressing this issue, and shows in details their drawbacks. Then, a new methodology of dynamic power estimation in TLM is introduced. The characterization process for power and TLM implementation is presented in the book. Finally, experimental results are demonstrated for the proposed methodology.

The book is divided into six chapters as follows:

- Chapter 1: This chapter is the introduction to the book. It starts with research scope about power estimation in TLM. Then, it illustrates the objectives of the research. First objective is to explore previous work in this area, and then propose and implement a new methodology. Finally, it presents the contribution of this book.
- Chapter 2: The chapter is exploring the research background. The needed fundamental concepts in this book are illustrated.
- Chapter 3: The chapter is presenting the previous work in power estimation in TLM and the drawbacks in the other techniques.

- Chapter 4: In this chapter, a novel methodology TLPM is proposed. The details of the characterization process for design analysis and power characterization are discussed. Building power models are illustrated. Finally, implementation and simulation of TLPM are presented.
- Chapter 5: This chapter is exploring the experimental results including the accuracy and time information for the proposed methodology.
- Chapter 6: The main conclusions of the work are provided in this chapter also potential future work.

References

1. Chandrakasan, A. P., Sheng, S., & Brodersen, R. W. (1992). Low-power CMOS digital design. *IEICE Transactions on Electronics, 75*(4), 371–382.
2. Mutoh, S. I., Douseki, T., Matsuya, Y., Aoki, T., Shigematsu, S., & Yamada, J. (1995). 1-V power supply high-speed digital circuit technology with multithreshold-voltage CMOS. *IEEE Journal of Solid-State circuits, 30*(8), 847–854.
3. Gonzalez, R., Gordon, B. M., & Horowitz, M. A. (1997). Supply and threshold voltage scaling for low power CMOS. *IEEE Journal of Solid-State Circuits, 32*(8), 1210–1216.
4. Borkar, S. (1999). Design challenges of technology scaling. *IEEE micro, 19*(4), 23–29.
5. Accellera Systems Initiative. Transaction Level Modeling. http://www.accellera.org.
6. Rose, A., Swan, S., Pierce, J., & Fernandez, J. M. (2005). Transaction level modeling in systemC. *Open SystemC Initiative.*
7. Accellera Systems Initiative. IEEE 1666-2011: SystemC Language Reference Manual. http://www.accellera.org.
8. Maillet-Contoz, L., & Ghenassia, F. (2005). Transaction level modeling-An abstraction beyond RTL. *Transaction level modeling with SystemC-TLM concepts and applications for embedded systems* (p. 23). Berlin: Springer.
9. Pasricha, S. (2002). Transaction level modeling of SoC with SystemC 2.0. In *Synopsys User G roup Conference (SNUG)* (vol. 3, pp. 3).
10. Brodersen, R., Chandrakasan, A., & Sheng, S. (1992). Low-power signal processing systems. In *Workshop on VLSI Signal Processing* (pp. 3–13).
11. Guyot, A., & Abou-Samara, S. (1998). Power consumption in digital circuits. In *International Conference on ASIC, Beijing China* (pp. 20–23).
12. Bellaouar, A., & Elmasry, M. (2012). *Low-power digital VLSI design: circuits and systems.* Berlin: Springer Science and Business Media.
13. Palnitkar, S. (2003). *Verilog HDL, a guide to digital design and synthesis.* Upper Saddle River: Prentice Hall Professional.
14. Sokolov, S. A. (2005). RTL power analysis using gate-level cell power models. Sequence Design, Inc., U.S. Patent 6,901,565.
15. Bombieri, N., Fummi, F., Pravadelli, G., & Marques-Silva, J. (2005). Towards equivalence checking between TLM and RTL models. In *Proceedings IEEE International Conference on Formal Methods and Models for Codesign* (pp. 113–122).
16. Ghenassia, F. (2005). *Transaction-level modeling with SystemC.* Dordrecht, The Netherlands: Springer.
17. Pasricha, S., Dutt, N., & Ben-Romdhane, M. (2004). Extending the transaction level modeling approach for fast communication architecture exploration. In *Proceedings of the 41st annual Design Automation Conference* (pp. 113–118).
18. Gajski, D. D., Zhu, J., Domer, R., Gerstlauer, A., & Zhao, S. (2012). *SpecC: Specification language and methodology.* Berlin: Springer Science and Business Media.

19. Beltrame, G., Sciuto, D., & Silvano, C. (2007). Multi-accuracy power and performance transaction-level modeling. *Transactions on Computer-Aided Design of Integrated Circuits and Systems, 26*(10), 1830–1842.

Part I
Background

Chapter 2
Fundamental Concepts

2.1 Introduction

In this chapter, the fundamental concepts for this book are presented. At first, different techniques of design modeling are presented. This provides a simple comparison between different techniques with respect to performance and why modeling at higher level is needed at early stages of design. TLPM methodology is based on power modeling in TLM using SystemC. Thus, an overview of SystemC is described. It is the HDL used in the implementation of methodology. Then, TLM is discussed in order to give a full overview of the implementation scheme including terminology and building blocks.

Universal Verification Methodology (UVM) [1, 2] is used to verify the created model functionality versus real design in RTL. Full description for building blocks and classes is described.

TLPM power characterization stage uses design synthesis and switching activity interchange format (SAIF) file. In this chapter, design synthesis concept is presented in details. Also, SAIF file standard is summarized in this chapter.

2.2 Design Modeling

2.2.1 Introduction

SoC design creates new challenges in the industry. Many design modules are being added to the same chip including cores and peripherals. The evolution of the number of modules per design is boosting. Additionally, technology nodes are shrinking tremendously. Billions of gates are used in the same design. The number of transistors in microprocessors is shown in Fig. 2.1 [3] as an example. Thus, the advance in the complexity of electronic design is affecting the product design cycle. Size and power

© Springer Nature Switzerland AG 2020
A. B. Darwish et al., *Transaction-Level Power Modeling*,
https://doi.org/10.1007/978-3-030-24827-7_2

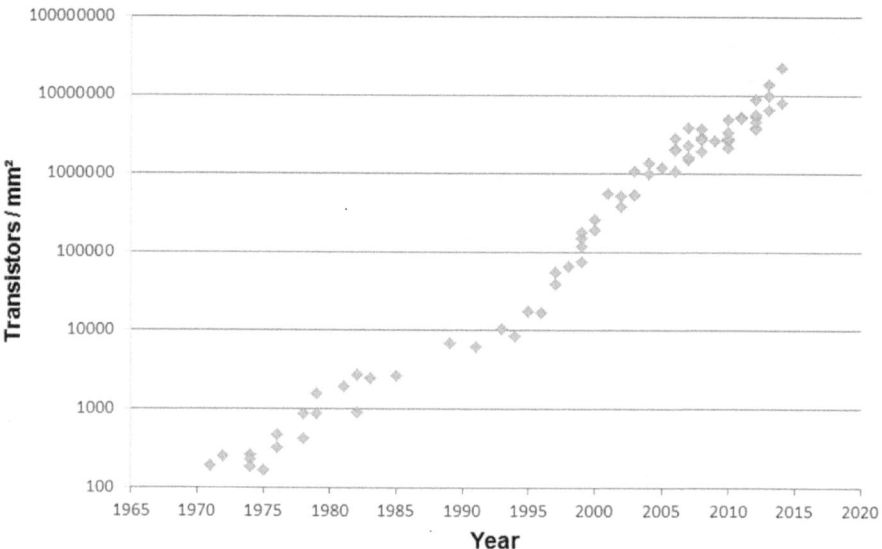

Fig. 2.1 Number of transistors in Intel microprocessors [3]

consumption of SoC should be kept at minimal values in order to reach the maximum performance at the least cost.

Modeling of SoC designs at high level of abstraction eases the complexity challenges. Design engineers can validate the functionality and performance at initial phases of the product design cycle, as abstraction at a higher level hides the design details at early stage. Hence, development and simulation speed of the abstracted models are much faster.

2.2.2 Design Abstraction Levels

Different levels of abstraction are presented in Fig. 2.2 [4]. Transaction-Level Modeling (TLM) [5] introduces abstract modeling of communication schemes between design modules using function calls. The implementation of modules is separated from the details of module communication in the design.

In RTL, a behavioral model with sufficient design details is created. Modules and interfaces constitute the fundamental design elements. Those modules and interfaces enclose interconnects, registers, and processes.

When the design is synthesized, variables and wires in a given RTL are translated into gates and nets representing the gate level. Every assignment in the design is mapped to combinational and sequential logic gates. Transistors are grouped in certain configuration to constitute the gates. The lowest circuit level represents the transistors and their interconnect models. The interconnect models are derived from parasitics extraction of the design after placement and routing.

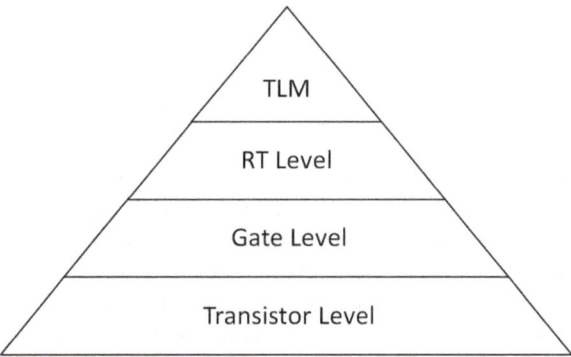

Fig. 2.2 Different levels of design modeling [4]

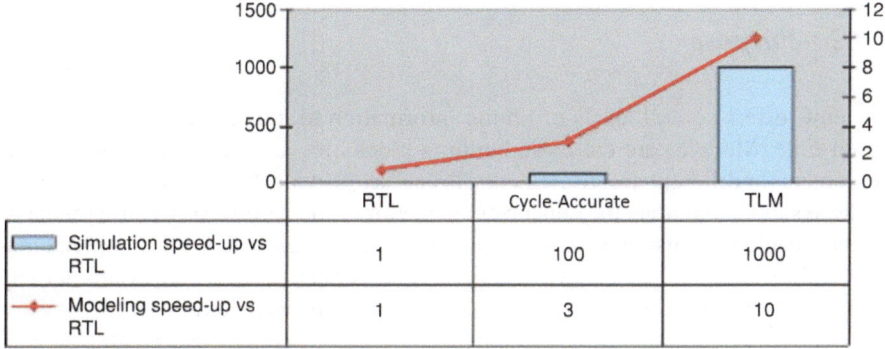

	RTL	Cycle-Accurate	TLM
Simulation speed-up vs RTL	1	100	1000
Modeling speed-up vs RTL	1	3	10

Fig. 2.3 Efficiency of modeling strategies [8]

2.2.3 TLM Design Abstraction

The conventional methodologies are not scaling well to cope with the increase in complexity of SoCs industry. RTL simulation is very computational and time-consuming for large designs [6]. Different methodologies and modeling techniques have been developed to keep the design life cycle short through different variants. Different layers of design abstraction are addressed in order to enhance productivity. TLM has been developed over many years making it good candidate for speeding up the design process, as shown in Fig. 2.3 [7].

TLM is characterized by simple and fast implementation of models. Also, simulation speed is much higher than RTL, as TLM holds abstracted design information. It is also possible to add timing information on the untimed models for analysis without the need for cycle-accurate (CA) models at early stages of design. Thus, TLM is introduced as an ideal complement to RTL when it comes to the trade-off between simulation speed and accuracy.

2.3 SystemC

2.3.1 Scope

SystemC is an HDL based on C++ class library [8]. It is used for system-level design to complement conventional HDL used in RTL as VHDL [9] or SystemVerilog [10]. It gives the capability for designers to create hybrid systems between software and hardware. SystemC provides different levels of design abstraction, which are needed for modeling of large SoCs, in order to have efficient and fast simulation [11]. This is achieved by exploiting object-oriented programming (OOP) in C++. It facilitates the creation of the needed objects for data communication across the design components.

2.3.2 Purpose

SystemC offers several levels of timing information to tune accuracy versus simulation time. Modules are the basic building blocks for a system. They are used as containers for different design elements. Processes provide the functional description of the system and imitate the concurrent behavior of the hardware. The use model of system-leveldesign using SystemC is to verify the design before RTL implementation, as illustrated in Fig. 2.4 [8]. Thus, the designer can catch severe issues at early stages of the product cycle minimizing the cost of expensive iterations.

2.3.3 SystemC Example

In this section, native SystemC classes or objects are mentioned below. Those classes have similar representation in other HDL, i.e., SystemVerilog. An example of adder using SystemC is illustrated in Fig. 2.5.

- Module is represented in SystemC by using class **sc_module**.
- While for ports, **sc_port** is used in the design. There are directions as well: **sc_in** and **sc_out**.
- An interface is derived from the class **sc_interface**.
- Meanwhile, event/signal are represented by **sc_event/sc_signal**, respectively.
- The main program of design execution is **sc_main**.
- The constructor for base module is performed by **SC_CTOR**.
- For functions, **SC_METHOD** is used.

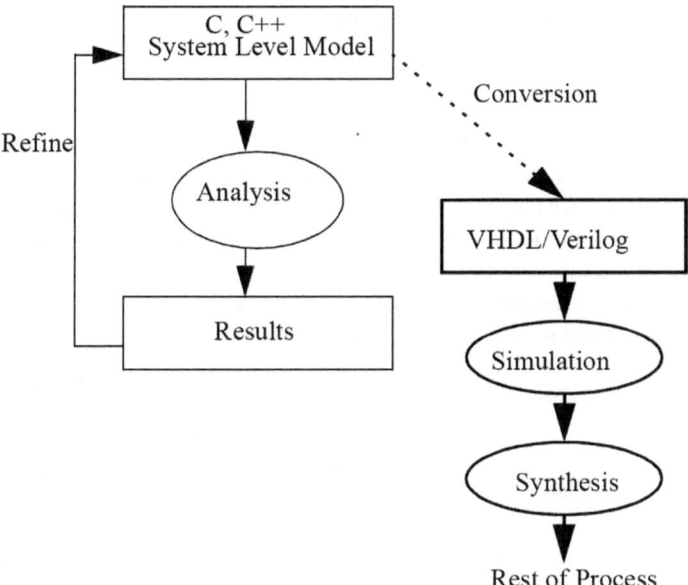

Fig. 2.4 System-level design cycle [7]

Fig. 2.5 SystemC example

```
#include "systemc.h"

SC_MODULE(simple_adder) {
sc_in<sc_uint> a, b;
sc_out<sc_uint> sum;

void do_add() {
    sum.write(a.read() + b.read());
}

SC_CTOR(simple_adder)          {
    SC_METHOD(do_add);
    sensitive << a << b;
}
};
```

2.4   Transaction-Level Modeling

2.4.1   Overview

TLM is an abstract modeling of communication schemes among design modules. SoCs are built in TLM by creating modules and channels and integrating them together. Channels are modeled to handle the communication across modules including the core and peripherals. Functions are executed through calls from the interfaces.

A "transaction" is the set of data being exchanged across ports. Data is exchanged between channels and modules through communication ports in order to perform the expected design behavior. An initiator module initiates transactions in the design, where target modules serve transactions.

2.4.1.1   TLM Classes

TLM is based on SystemC library as described in Fig. 2.6 [8]. TLM-2.0 is current version standard, which based on old TLM-1.0 classes and newly introduced interfaces and utilities. Here is a list of new support in TLM-2.0:

1. Group of core interfaces (blocking transport interface, non-blocking transport interface, direct memory interface (DMI), and debug transport interface).
2. Initiator and target sockets,
3. Generic payload for modeling of memory buses, and
4. Global quantum.

2.4.1.2   TLM Use Cases

TLM supports different usages in the industry. Every use case differs in the used interface and the corresponding timing model (coding style).

An algorithmic model which is based on single thread in software. Thus, it is not considered genuine TLM model, because TLM is communication scheme and cannot be maintained by single process.

The ideal abstract model should contain multiple processes (threads) to maintain communication. This needs special mechanism to handle control across process. Synchronization between the sequence of events should be maintained, in order to define strong one. Strong synchronization is implemented using semaphores or FIFOs. This allows complete simulation of untimed model without advance in time.

Approximately timed model should associate multiple protocol specific timing points for each transaction, in order to mark the start and the end of each phase protocol. A complete list of use cases is listed in Fig. 2.7 [8].

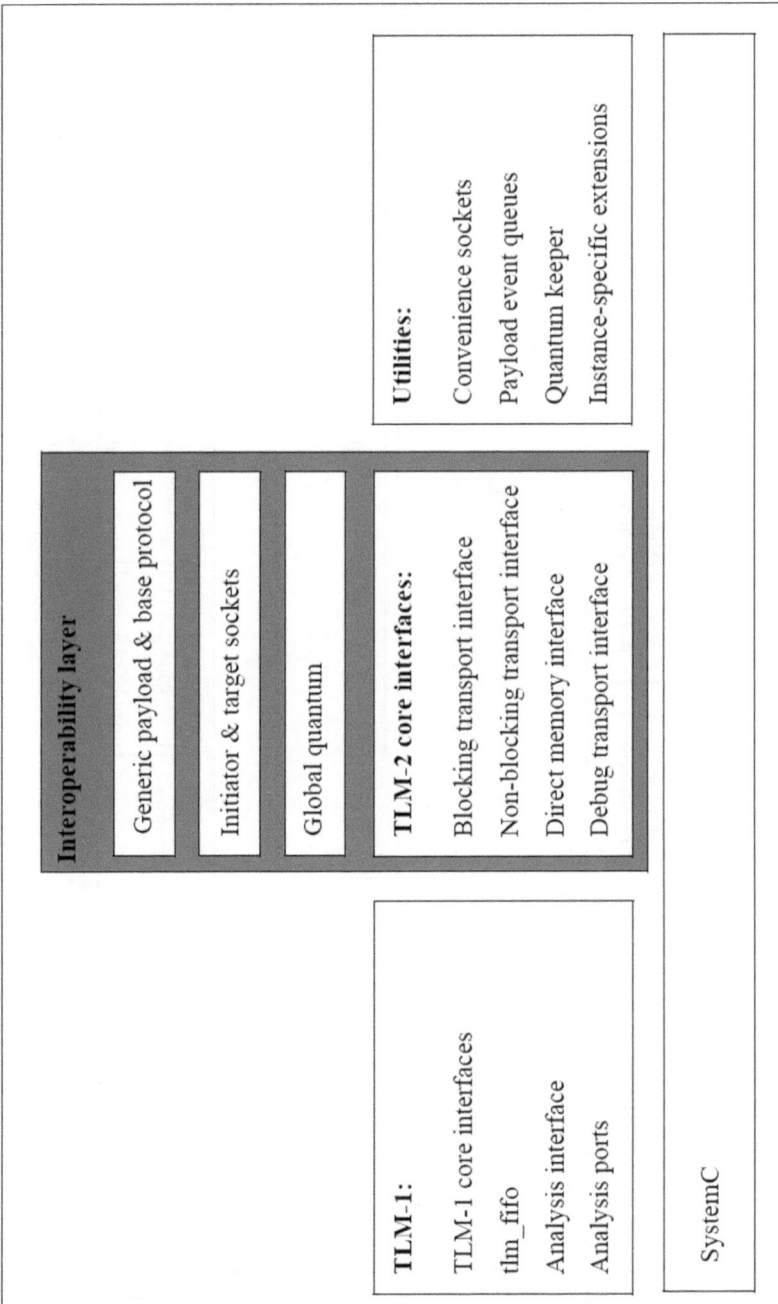

Fig. 2.6 TLM classes [7]

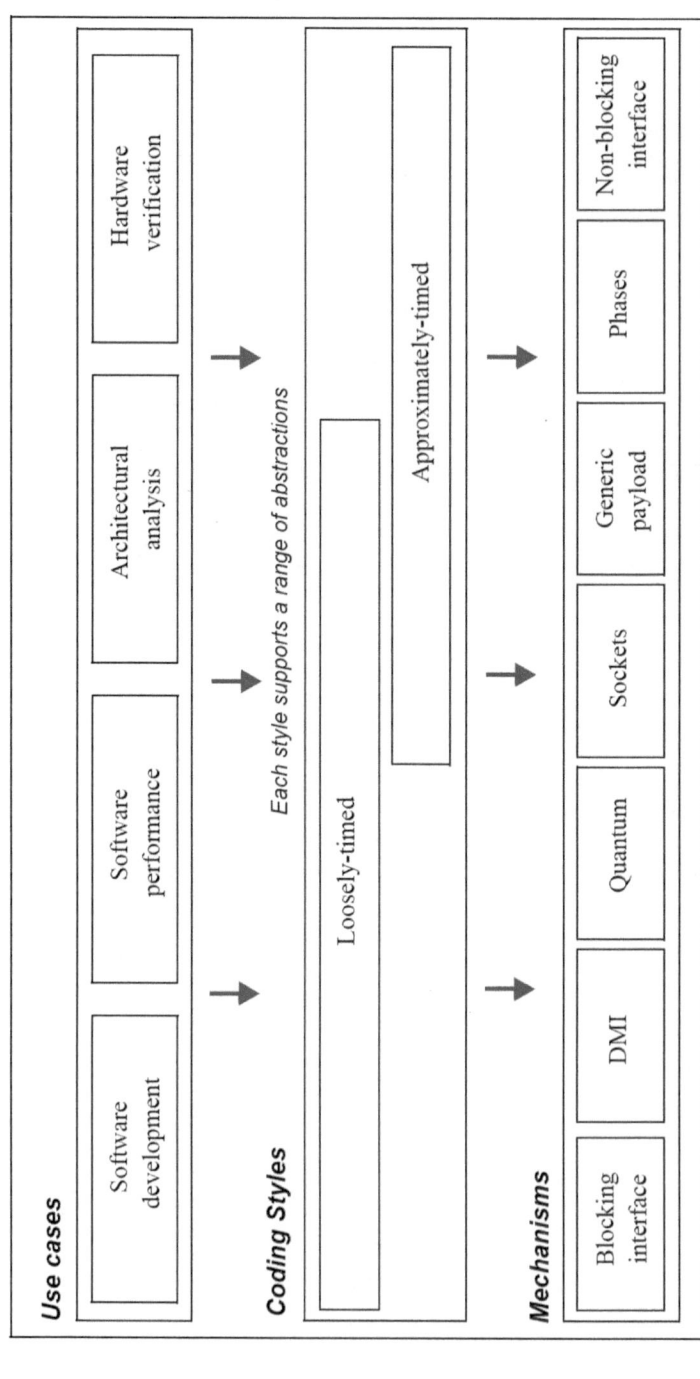

Fig. 2.7 TLM use cases [7]

2.4.1.3   Coding Styles in TLM

A TLM user implements the code in different algorithms. Each one represents group of programming semantics that are suitable together. It should not use specific interface or abstraction level; they are independent of the coding style. The conventional semantics used by designers are listed as follows.

Untimed Coding Style

As mentioned before, untimed models are not a part of TLM-2.0, because any bus-based systems need timing information representing real software running on processor.

Loosely Timed Coding Style and Temporal Decoupling

The interface that is used in loosely timed style is blocking transport interface. It allows couple of timing information for each transaction. The allowed information are the call for of the blocking transport function representing the start of request and the return from this function which represents the response. The request and the response could be at the same simulation time or at different times according to the design.

In that case, loosely timed style could be suitable for software development as presented in Fig. 2.7. Designers go for virtual platform model where the software include one or more operating system. Loosely time style is able to boot the system and run code on target machine, due to modeling support of timers and interrupts.

On the other hand, loosely timed supports temporal decoupling. Temporal decoupling means that single process can advance in time (a local one) without advancing in the whole-system simulation time. The single process advances in time till reach a point where synchronization is a must versus other processes in the system. Thus, it would improve the simulation wall time tremendously for many systems as it increases the locality of data and reduces the overhead of scheduling across the system. As a conclusion, each process is allowed to run for quantum (time portion) before advancing the whole-system simulation time.

Regarding the scheduler itself, it advances in simulation time to reach the following event, when this event has any running-related processes sensitive to that event. The processes are only running in the current simulation time, whenever process read/write variable, the information is only related to the current simulation time. The advantage of running in the local rather than continuous synchronization with kernel; that the overhead for communication with simulation kernel is kept to the minimum. So, no need for frequent back/forth with kernel (specially, this communication is considered the dominant factor in simulation time).

Processes are responsible for determining if it can advance the simulation with local evaluation or not. This main requirement in that case is to keep functionality intact. When the process finds an external dependency, it goes for one of two paths:

1. Forcing Synchronization:
   The process hands over the control to simulation kernel; this allows other processes to catch-up this advanced process till the simulation time arrives at the

needed time. This complies with the semantics of SystemC and will keep functionality harmony.

2. Sample/update the signal and continue the process:
This is based on an assumption that the communication and timing have certain expected setup. Thus sampling or updating values early or late do not affect the functionality. This is valid in the proposed use case of virtual platform simulation, as the software stack has nothing to do with the lower details in hardware.

However, there is a precaution for such implementation. If the process is permitted to run local without synchronization with the system kernel, the scheduler would not be able to catch many changes that have been done. Also, other process in the system would not operate. Thus, this is avoided by adding "global quantum." It represents upper limit for the process to advance than simulation kernel. Quantum is the trade-off between simulation speed and accuracy in functionality. When the quantum is too large, this may introduce inconsistency in timing across the system and result in loss in functionality. On the other hand, if the value is too small, it synchronizes frequently with the simulation kernel and downgrades the simulation speed. Quantum is defined by the application itself.

Temporal Decoupling Example

Considering system contains the following: processor, memory, timer, and external peripherals (would contribute with slowness in the system). Software of the processor spends most of time executing instructions through communication with memory module. This operation is paused for further communication with the system in case of interrupt from the timer, i.e., 10 ms. Thus, the preferred quantum for this system is 10 ms. The processor communicates with memory for long time, till the quantum is reached. Then, it starts communication/synchronization with other modules (peripherals, which are considered bottleneck for simulation speed in the system). In this system, software is isolated from the activity of hardware; so it reaches high speeds 100X without losing the coherency of functionality.

In general, some process in the system could be potential target for temporal decoupling and others are not. As well, quantum value could be different across them. This is based on the nature of the process and its dependency.

Loosely Timed Coding Style Synchronization

The untimed model depends on the availability of synchronization points to pass control between process and simulation kernel. However, for loosely timed models, functionality could be kept intact by quantum usage across processes. Thus, every process can advance in execution before yielding the control back to the mainstream. And accuracy could be improved by embedding explicit synchronization in the system. On very high level of system, explicit synchronization is needed.

Approximately Timed Coding Style

Non-blocking transport interface supports the approximately timed style. This could be used for architectural/performance analysis. Because it provides many points for timing and phases for every transaction.

Each transaction is composed of phases, where each phase has the corresponding timing points for transition between phases. For a base protocol, there are the request and the response. Thus, there are four timing points for the start/end of each of them. Complex protocols may need more timing points but that will affect the compatibility with generic payload.

Approximately timed coding style does not use temporal decoupling concept, as there is a need for high accuracy for timing. The process executes in lock step with the simulation kernel, while interactions are annotated with delays.

In order to achieve this model, the delays are annotated in order to represent the data transfer times for commands (read/write) latency of the target. The data transfer time is the same as the minimum acceptable delay between two successive requests/responses. The delays are implemented by interaction with SystemC side using "wait" or "notify" statements.

Non-blocking transport interface could be used as well for loosely timed models using just two timing points. However, the blocking transport interface is more preferable.

Differences Between Coding Styles

1. Loosely timed:

   - Transaction has two timing points.
   - Simulation time is considered.
   - Processes have the utility of temporarily decoupling.
   - Process is responsible for tracking its advance with respect to simulation kernel.
   - The process could yield the control to kernel in two cases:
     (a) Explicit synchronization point.
     (b) Time quantum consumption.

2. Approximately timed:

   - Transaction has many timing points.
   - Simulation time is considered.
   - Processes run in lockstep with simulation time along with SystemC.

3. Untimed:

   - Simulation time is not considered.
   - Processes have predetermined synchronization points to yield.

Switching Between Approximately Timed and Loosely Timed

The design can be modeled to switch between approximately timed and loosely timed, in order to have compromise between accuracy and speed. As, the model can run reset/boor sequence rapidly at loosely timed, while it switches to approximately timed for more detailed analysis at critical stages of simulation.

Cycle-Accurate Modeling

Cycle-accurate modeling is not part of TLM-2.0 standard; however, it could be implemented by SystemC and TLM-2.0 using approximately timed coding style.

Blocking Versus Non-blocking Transport Interfaces

Any transport should support both blocking and non-blocking functionalities for timing detail. The blocking interface is capable of modeling the start and end of a transaction. In this case, the transaction is completed in a single function call. On the other hand, non-blocking transaction supports the breaking of transaction into many timing points. Thus, many function calls are needed for a single transaction.

The model that creates transactions could use blocking or non-blocking transport interfaces or even both according to the used coding style. Thus, it supports interoperability feature for the interface.

TLM provides convenience socket to convert incoming transport calls for both blocking and non-blocking into transport calls, which has some cost especially for converting incoming non-blocking call into blocking one. But this not a concerned drawback, as approximately timed model is a bottleneck in most of cases.

The static typing of C++ forces the implementation of both interfaces; however, initiator chooses to call either blocking or non-blocking dynamically. Also, initiator can switch between calls of blocking and non-blocking.

In general, the blocking transport interface is being used for simple coding style for complete transaction within single call, while for non-blocking transport interface supports the availability of many timing points for single transaction. Either type of interface supports temporal decoupling, which could be affected in negative way when using many timing points in approximately timed models.

Coding Styles and Use Cases

1. Loosely timed:

    (a) Software application development.
    (b) Software performance analysis.
    (c) Hardware architectural analysis.
    (d) Hardware functional verification.

2. Approximately timed:

    (a) Hardware architectural analysis.
    (b) Hardware performance verification.
    (c) Hardware functional verification.

3. Untimed:

    (a) Hardware functional verification.

4. Cycle accurate:

    (a) Hardware performance verification.

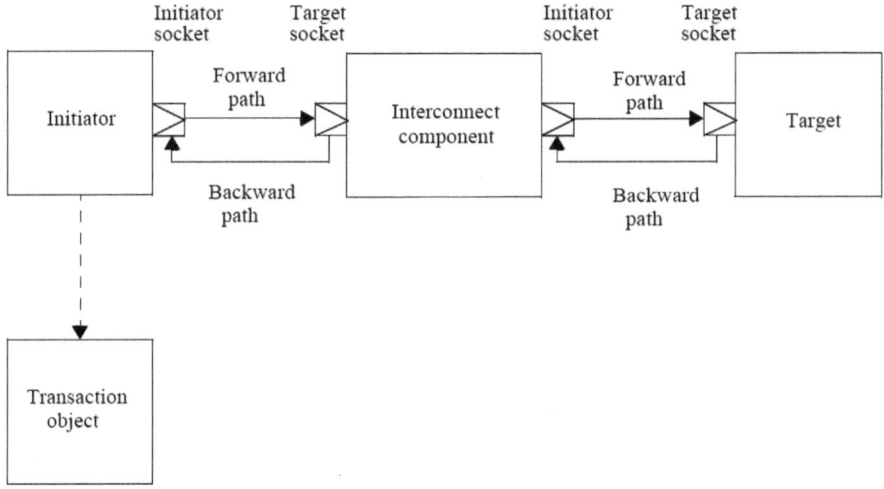

Fig. 2.8 TLM components [5]

2.4.1.4 TLM Components

TLM transfers the data across different design modules. This process is executed by function calls added to design registers. The communication is done through module sockets, interfaces, and bus models. TLM sockets facilitate the communication in both forward and backward directions. It also supports various protocols of bus models [12] to satisfy the needs of the system-level designer. Simple representation of TLM structure is presented in Fig. 2.8 [5].

Initiator is the module that creates the transaction and sends it by calling core interface method. Target is the final destination module for a transaction, such that in case of write transaction, initiator (i.e., processor) creates the data to be passed to target (i.e., memory). However, in case of read transaction, initiator receives data from target. The medium of communication between them is interconnect component (i.e., arbiter or router), which is neither initiator nor target.

These roles are not hard-coded in the model; they can switch dynamically throughout the process depending on the functionality of transaction. For example, a component can act as target for transaction but it is an interconnect for another one.

Transaction Lifetime Example

The object of transaction is created by initiator component. The passing mechanism is handled by transport interface either blocking or non-blocking, such that the object is passed as an argument for a method. This method is implemented using interconnect component, i.e., arbiter. The interconnect reads the attributes of the transaction before sending it to another transport call. The second call is also implemented by another interconnect, i.e., router. The second interconnect sends the object to third transport

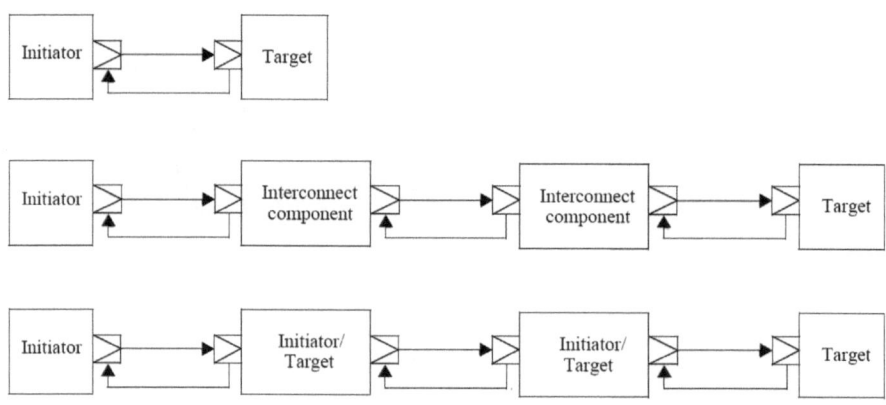

Fig. 2.9 TLM initiator, interconnect, and target [5]

call which is the target of the transaction. The number of interconnects could vary from zero to many based on transaction functionality. This is illustrated in 2.9 [5].

The mentioned sequences of method calls are called "forward path," as the transaction information is being passed from initiator component to target component. However, when the transaction is processed in target and it would be then transferred to initiator, it is performed in two ways.

1. Return path:

   - The information is carried with the return of method calls of transport interface as they unwind.

2. Backward path:

   - The information is carried out by making explicit transport method calls.

Thus, forward path is the direction where initiator or interconnect components makes method calls forward in direction of the target or another interconnect. On the other hand, backward path is the direction where target or interconnect components makes method calls backward in direction of initiator or interconnect. The path between initiator and target is composed of set of hops. A hop is the connection between two neighboring components, so number of hops is greater by one than the number of interconnects between initiator and target. When the generic load is used in model, both paths (forward and backward) should pass through the same number of components/sockets.

Each connection between components needs port and export, in order to support forward and backward paths. Port and export should be bound for compatibility, which is facilitated by sockets (initiator/target). Initiator socket is needed as a port for forward path for interface method calls and as an export for backward path. It is the opposite for target socket. Those sockets can encapsulate DMI and debug transport interfaces.

On using sockets, initiator should have at least one initiator socket, and target should have at least one target socket. While for interconnect component, it should have at least one each socket type. Components could have many sockets based on the transactions that they handle.

Bus Bridge Example

The designer can model the bus bridge in two ways:

1. Interconnect component:

    - The component passes transaction object as a pointer.
    - This speeds up the simulation time.

2. Transaction bridge between two separate transactions:

    - It requires copying transaction information.
    - It provides flexibility as the two transactions could have different attributes.

The usage of sockets is needed for maximum consistency, interoperability, and convenience.

2.4.1.5 DMI Interface

The Direct Memory Access (DMI) interface is different from transport interface. It is specialized interface to provide direct access to part of memory that is owned by target. When DMI has the access to the memory pool, it enables the initiator to avoid the conventional path using interconnect components as in the transport interface. Thus, it accelerates the memory transaction in loosely timed simulation. DMI supports both forward and backward paths.

2.4.1.6 Debug Transport Interface

Debug transport interface is also different from conventional transport interface. It is specialized to give debug access to memory that is owned by target. This access is very fast, as it is free from delays or conventional transactions side effects. However, debug transport supports only forward path.

2.4.1.7 TLM Interfaces/Sockets Combination

The transport interface with its both types (blocking/non-blocking) can be combined with DMI and the debug transport interfaces in sockets of initiator and target. They can be used as parallel to access information at a given target. These interfaces are the keys to

1. Ensure interoperability between components and
2. Being generic payload.

The initiator/target sockets in the standard provide the four interfaces. It implements the methods needed for the four of them. However, it is recommended to choose either between blocking or non-blocking transport interface used in socket according to accuracy/speed needs.

The coding style guides the choice of which interfaces to be used; so typically in the design, they are used as follows:

- Loosely timed style:

  - Blocking transport interface,
  - DMI, and
  - Debug transport interface.

- Approximately timed style:

  - Non-blocking transport interface and
  - Debug transport interface.

2.4.1.8   TLM Interfaces Summary

TLM provides different types of interfaces for different purposes. They are listed as follows:

1. Transport Interface:

   - It is the primary interface type for transaction communication between design initiator and target modules.
   - There are two types of transport interface: blocking and non-blocking transport interface.
   - Both types support the annotation of timing information.

2. Direct Memory Interface:

   - This type is much faster than normal transport interface.
   - This behavior accelerates the simulation by avoiding the normal interface operation and access the memory using pointers.
   - There are two types of DMI to serve the forward and backward paths of communication.

3. Debug Transport Interface:

   - This type of interface is used for debugging purposes only.
   - It works in the forward path only and it does not support annotation of timing information.

2.4.1.9 Levels of Abstraction in TLM

TLM classifies the models according to various characteristics. Those characteristics include timing information, model evaluation frequency, and the level of abstraction in computations and communications. Designers create models at different levels of abstraction. Each level of abstraction represents an implementation scheme for design components and timing information. Those abstraction schemes vary according to the desired the accuracy of models and transactions. Thus, there is a trade-off between accuracy and speed for different abstraction levels and the corresponding timing information.

For timing information, the model could be untimed, approximately timed, and cycle accurate [13]. Untimed model is mainly used for very early verification. Although this model does not have any timing information or clocks, it must perform its functionality correctly for data handling across processes. However, untimed model cannot capture detailed architectural information, like timed TLM models. For timed models, complete detailed information about delays and synchronization are provided. The delays are classified as computational delay and communication delay. Computational delay is the time needed to calculate certain functionality inside a module, where communication delay is the time needed to handle and transfer data across initiator and target modules. At the same time, the data representation is categorized into three levels: full information of pin level, bus packet, and application packet.

TLM has different levels of abstraction as illustrated in Fig. 2.10 [7]. When the model is untimed and data is transferred as application packet, the model is called

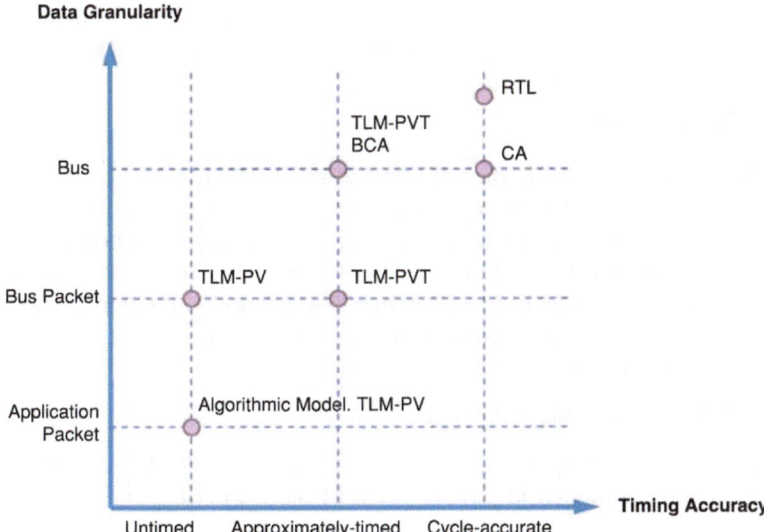

Fig. 2.10 TLM levels of abstraction [8]

algorithmic model. Upon enhancing the model using bus packet communication or timed modules, the model is called a programmer view (PV) model. If it includes both bus packet and timing information, the model is called a programmer view plus timing (PVT). For detailed bus and accurate timing information, the model cycle-accurate models till implementation complete in RTL.

2.4.2   TLM 2.0 Interfaces

In the part, the details of implementation of needed interfaces for TLM-20.0 modeling are blocking/non-blocking transport interfaces, DMI, and debug transport interface.

2.4.2.1   Transport Interfaces

The transport interfaces are the primary ones for communication of transactions between different TLM components (initiator/target/interconnect). Both versions of transport interface (blocking/non-blocking) support temporal decoupling and timing information annotation. But only non-blocking transport interface supports the availability of multiple phases in transaction lifetime, while blocking transport interface does not have this feature as it does not have any explicit phase argument. Only non-blocking methods have a return argument for the value of return path.

In order to have fast, abstract modeling for memory mapped buses, the transport interfaces, and generic payloads are intended to be designed together. However, the templates of transaction interfaces give the capability of standalone from generic payloads to support different purposes. The implemented rules for memory management, transaction ordering, and method calls sequence rely on transaction types based on the used protocol traits.

Blocking Transport Interfaces

The interface supports loosely timed coding style. This style is convenient when the transaction is completed within single function call. The transaction timing points are just two: the start and the end of the transaction. The blocking transport interface only supports forward path for communication between initiator and target. The interface has b_transport method that has two arguments which are transaction argument and timing annotation argument. Different scenarios for transaction communication are listed in the following timing charts.

Message Chart—Blocking Transport

b_transport method can return at different cases either immediately or yield control to scheduler and return later in simulation, as illustrated in Fig. 2.11 [5].

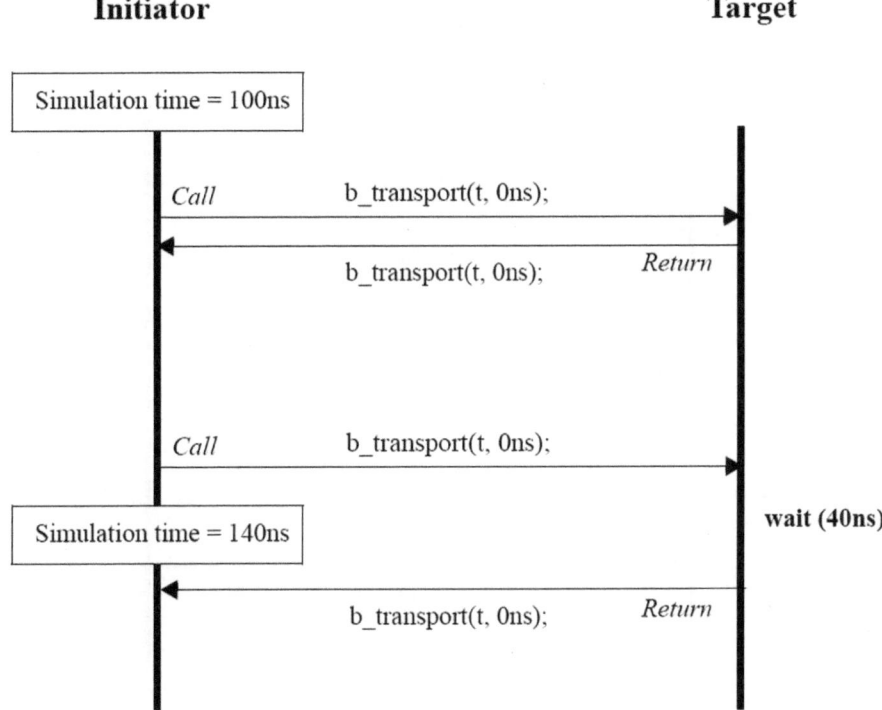

Fig. 2.11 Blocking transport [5]

Message Chart—Temporal Decoupling

Initiator can run in local time in advance of simulation time; thus, the nonzero value is passed to time information argument in b_transport method, as illustrated in Fig. 2.12 [5].

Message Chart—Time Quantum

When initiator runs in temporally decoupled mode, it could advance till the specified quantum. Then, it has to yield the control back to the main process, so that other initiators/processes can work and proceed in time. Thus, the main purpose of delays in loosely timed coding style is to give the model information about how much to proceed in local run before simulation. An example is illustrated in Fig. 2.13 [5].

Non-blocking Transport Interface

The non-blocking interface supports approximately timed coding style. It has the functionality to break down the transaction into multiple phases. Each phase has its corresponding timing point, where each call/return can correspond to a phase transition.

If the number of timing points is restricted to two, the non-blocking transport interface can be used in loosely timed coding style. But this is not recommended

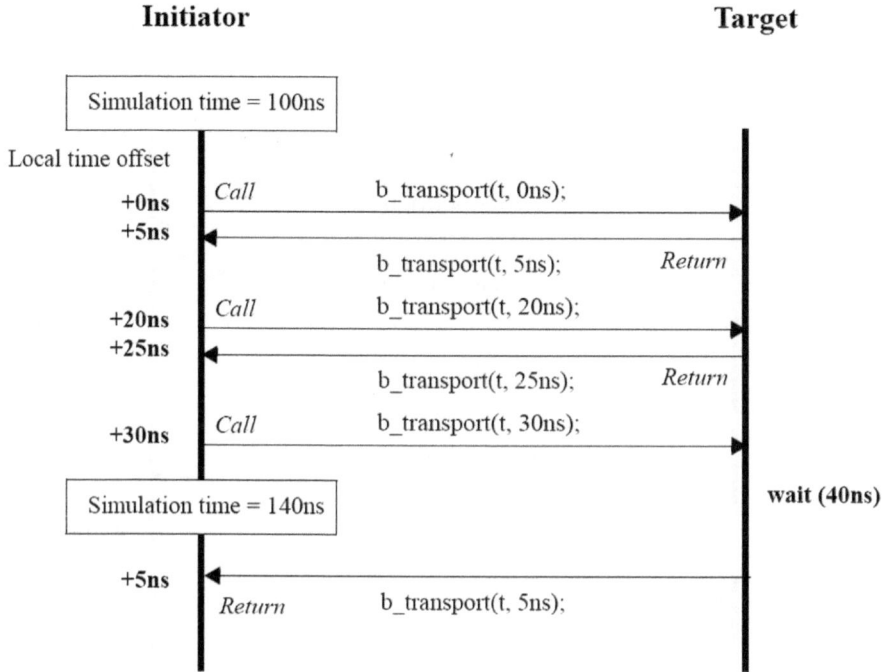

Fig. 2.12 Temporal decoupling [5]

due to the complexity of the non-blocking transport interface than blocking one. The non-blocking transport interface is intended for pipelined transactions.

It supports both forward/backward paths for transactions between initiator and target. There are two declarations for interface to support both directions. The interface method has three arguments which are transaction argument, timing annotation argument, and phase that indicates the transaction state (this item is extra than the blocking interface method). Different scenarios for transaction communication are listed in the following timing charts.

Message Sequence—Using Backward Path

Regarding the calling sequence for non-blocking interface, the arguments and return value are passed to interface. The phases of base protocol are used: BEGIN_REQ, END_REQ, BEGIN_RESP, and END_RESP. For approximately timed and base protocol, transaction is communicated back-and-forth couple of times between components. If the transaction is not calculated immediately, return value should be TLM_ACCEPTED. Thus, the caller would yield control to the main system. An example is illustrated in Fig. 2.14 [5].

Message Sequence—Using Return Path

If the receiver (i.e., target component) can expect immediately the behavior of transaction with respect to state and delay, then the receiver can return the new state

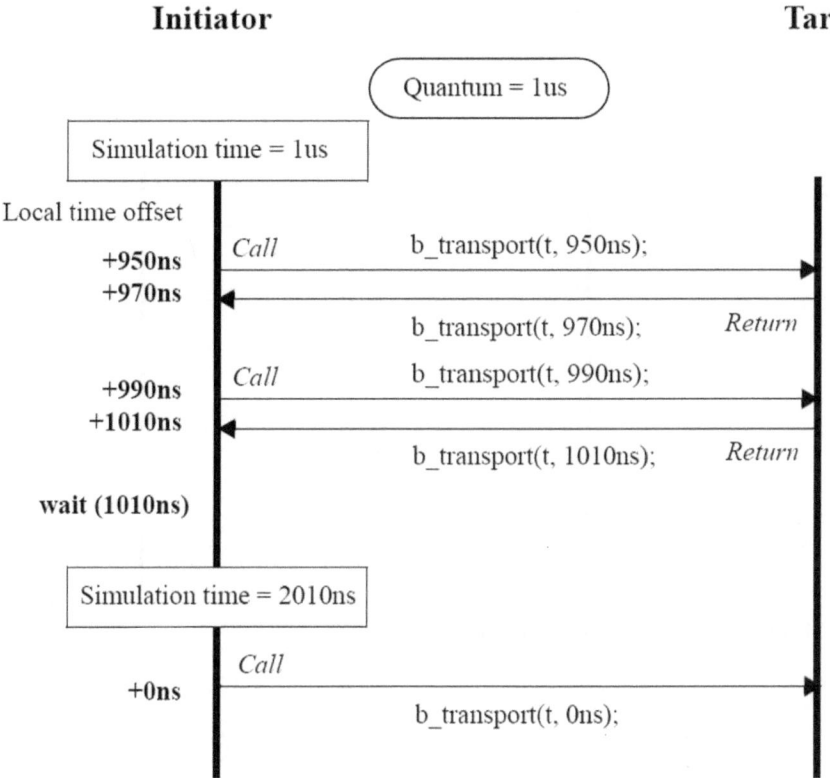

Fig. 2.13 Time quantum [5]

of transaction without the need of opposite path. The states that can be provided are TLM_UPDATED and TLM_COMPLETED. When TLM_UPDATED is issued, transaction, phase, and time information should be annotated. TLM_COMPLETED could be used at any stage if all transaction phases are already done. An example is illustrated in Fig. 2.15 [5].

Message Sequence—Early Completion

If a component (initiator/target) returns TLM_COMPLETED from the interface at early time for transaction completion, no further call is needed. In Fig. 2.16 [5], different phase transitions are implicit.

Message Sequence—Timing Annotation

Time delays could be annotated in calls to indicate when the transaction would be processed. This is handled in approximately timed coding style by adding the transaction into payload event queue. The processing is handled either in SystemC process or in a registered callback. The delays are annotated in calls in either direction

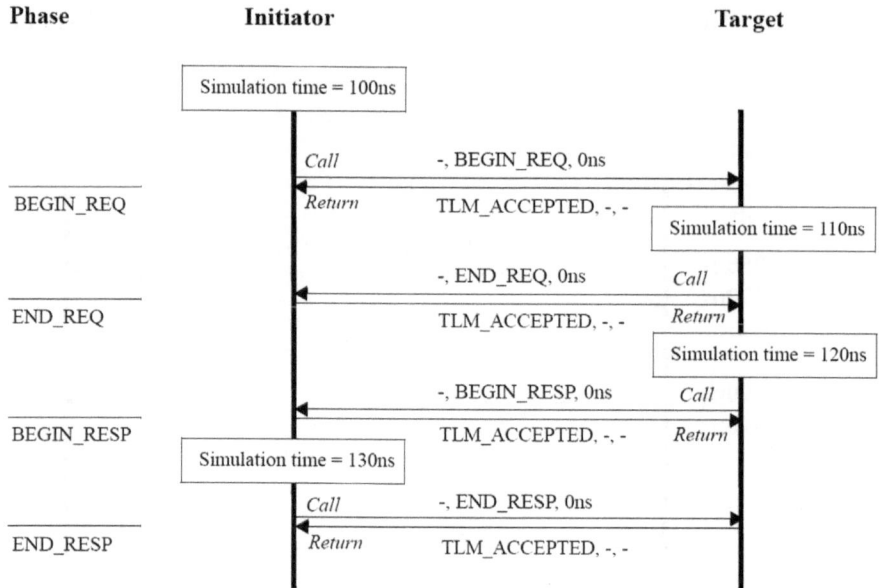

Fig. 2.14 Message sequence—backward path [5]

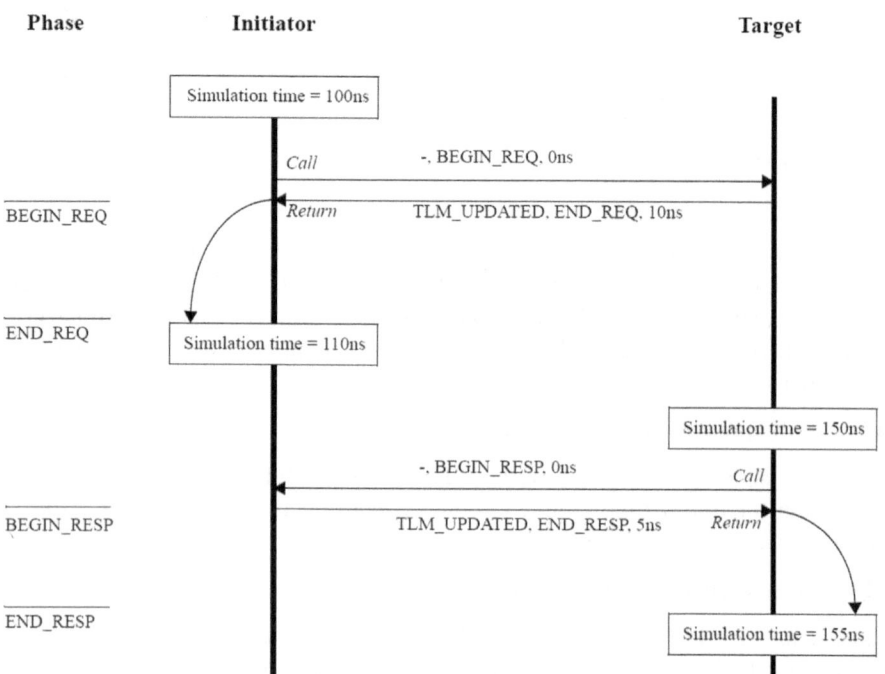

Fig. 2.15 Message sequence—return path [5]

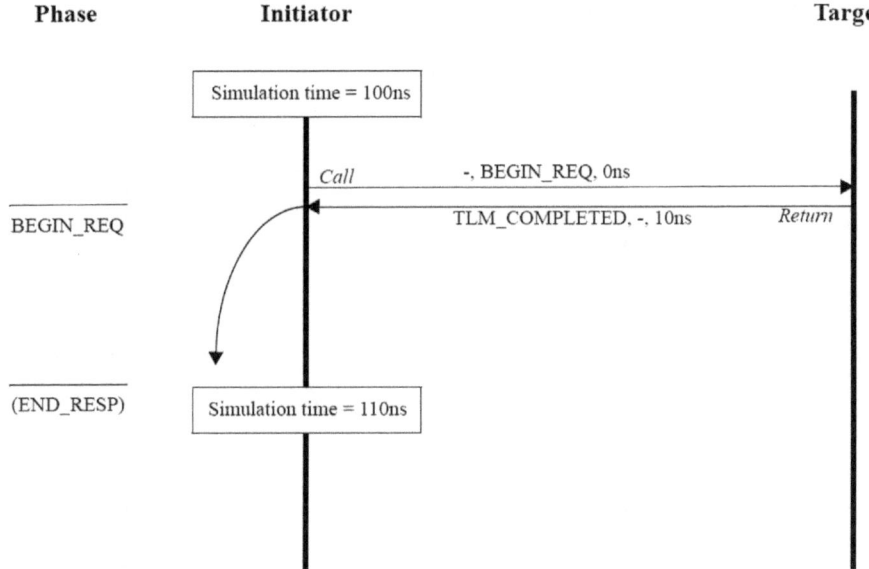

Fig. 2.16 Message sequence—early completion [5]

(forward/backward). An example is illustrated in Fig. 2.17 [5], and different phase transitions are implicit.

Timing Annotation Annotation of timing is a common feature for both blocking and non-blocking interfaces. It is added as argument with sc_time value for interfaces methods.

Usage Rules

1. Transaction object should have timing information.
2. The time argument is a positive value.
3. The time argument can be added to both call and return of interfaces methods.
4. For non-blocking interface methods, it can only increase the time value argument, while, for b_transport, it can increase or decrease the value.
5. When transactions are created by different initiators, thus according to temporal decoupling, the interface method calls order is different than local time order. The recipient is responsible for out-of-order transactions.
6. When a nonzero timing value is passed as an argument, the receiver can choose between speed or accuracy. This is based on the used coding style whether it is loosely timed or approximately timed.

    (a) In case of speed need, transaction can be processed immediately without any delay. The system is expected to deal with out-of-order processes.
    (b) In case of need of accurate model, the transaction should be processed in the correct time with respect to other SystemC processes.

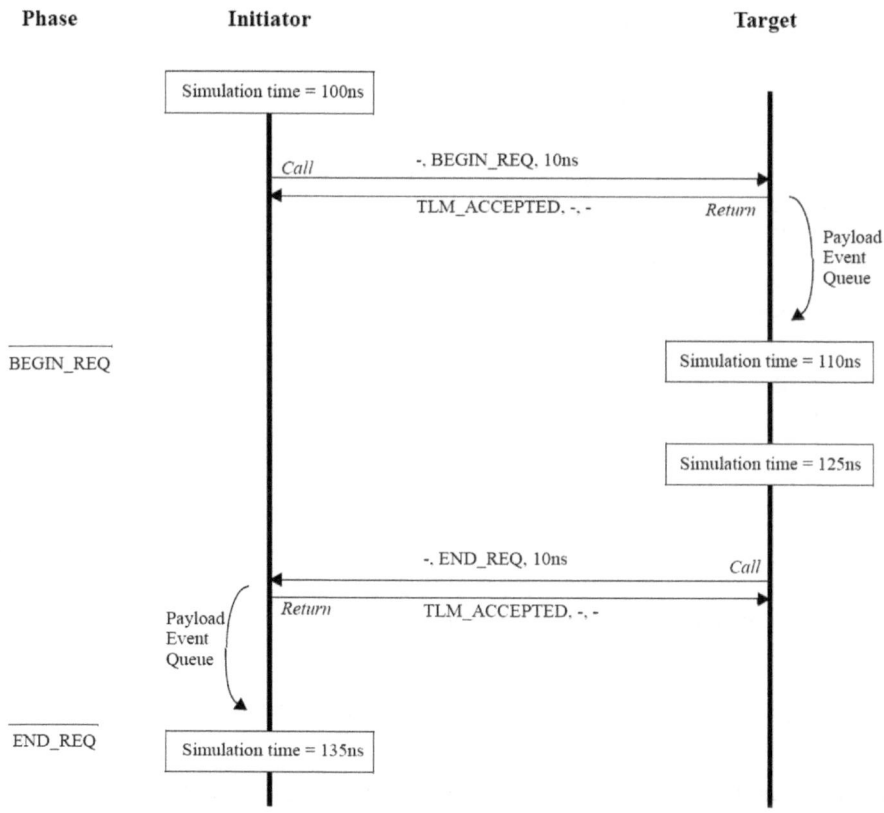

Fig. 2.17 Message sequence—timing annotation [5]

7. In loosely timed coding style, b_transport is the recommended method to use, in order to execute the transactions immediately.
8. In approximately timed coding style, the transactions are expected to have delays; thus, non-blocking interface methods are recommended for usage.
9. The components are expected to perform the choices dynamically, to perform the transaction immediately or with delays.
10. The component is free to choose the mutual execution order of method calls if they are out-of-order than local order execution.
11. Temporal decoupling can describe timing annotation.

2.4.2.2 Direct Memory Interface

Direct Memory Interface (DMI) provides the capability to directly access the memory information owned by target using a pointer. This capability gives high potential for accelerating the model simulation speed, as there is no need to go through the conventional interfaces for communication.

DMI has two interfaces, one for forward path from initiator to target and the other for backward path from target back to initiator. The forward path needs certain mode for of DMI access for read/write operations to a given address in the memory pool. The backward path uses the pointers from forward path. Both paths could go through zero or many interconnect components, under the condition that they are equal in number in both directions.

On passing transaction on forward path, a DMI pointer is requested. The arguments are the command and the address of the transaction. DMI provides the latency values of the model, so it gives sufficient information and accuracy for loosely timed coding style.

Transport Interface and DMI Comparison

1. Transport interface should go through interconnect component, while DMI can bypass it.
2. Correct behavior in interconnect components should be retained, as transport interface could update the state, while DMI is bypassing it. Thus, DMI access should be denied.
3. Same initiator or different ones could use both transport interfaces and DMI. Both interfaces differ in important specification, i.e., timing annotation. Thus, application is responsible for ensuring correct behavior and data communication.

2.4.2.3 Debug Transport Interface

The debug transport interface can read/write to the target storage using the same forward path of conventional transport interface between initiator and target. Thus, it can perform the same address mapping as the conventional interface. This is performed without any delays or wait statements or events.

The debug transport interface grants different capabilities to work with on the design. It can check the memory addresses. Also, it can read the memory contents or even initialize some area in system memory after elaboration.

The default transactions used in the interface is tlm_generic_payload. It has the following arguments: command, address, data length, and data pointer.

Usage Rules

1. The debug interface should follow the same forward path as conventional interface.
2. The initiator is responsible for creating/managing the transaction with the suitable attributes.
3. The initiator sets the command attribute according to the needed type of access.
4. The address attribute is the first address in the memory for read/write access.
5. The interconnect components interact and modify the attributes of debug transaction as already performed in transport transaction.

   (a) The address attribute could be modified several times on passing on different interconnect components.

6. The initiator can reuse transaction objects across different types of interfaces.
7. The debug interface cannot call "wait" statement.

2.4.3 TLM-2.0 Global Quantum

Time quantum is the amount of time which SystemC processes can run ahead in simulation. This is performed with the loosely timed coding style using temporal decoupling capability. There is no need for explicit synchronization between the processes, as the process can run ahead in simulation for time quantum. However, the processes would need global quantum.

The delays in transport interface methods represent the local time offsets relative to the current simulation time in case of temporal decoupling. The global quantum is the default value between successive quantum values.

2.4.4 Combined TLM-2.0 Interfaces

The group of TLM-2.0 interfaces (transport, DMI, and debug transport) is used for forward/backward paths with initiator and target sockets in the design. The forward path is handled by forward interfaces by providing calls from initiator sockets to target. While for backward path, it is handled by methods from target sockets to initiator ones. As mentioned before, blocking transport interface and debug transport interface cannot provide the backward path.

In order to create new sockets for method calls using the interfaces, it could be performed by two approaches:

1. Define new socket class and instantiate it using the combined interfaces; however, this is not recommended for interoperability.
2. Derive new socket class from the standard socket class.

The combined interface is parameterized using the protocol traits. This class defines for the system the types (payload/phase) used by the interfaces. The class is associated with certain protocol (i.e., tlm_base_protocol_types as the default one).

2.4.5 TLM-2.0 Sockets

The socket element is used to link a port with an export. For initiator component, the socket has port in forward path and export in backward one. On the other hand, for target component, the socket has export in backward path and port in the forward one. Port/export is intended to bind to export/port of the opposing socket.

The most used socket classes are tlm_initiator_socket and tlm_target_socket. The sockets are parameterized with the protocol traits for forward/backward interfaces. They are bound together if they have the same protocol type.

The benefits are from sockets' usage:

1. The initiator and target sockets group different interfaces for forward/backward paths.
2. They can bind port/export for both paths in single call using methods.
3. Checking the types of binding incompatible protocols.

2.4.6 TLM-2.0 Phases

Non-blocking interface uses tlm_phase class as default phase type. The object from this class represents the phase. The enumeration values for the assignment correspond to four phases:

1. BEGIN_REQ
2. END_REQ
3. BEGIN_RESP
4. END_RESP

For interoperability, these four phases should be only used. If further phases are needed in the system, user should use TLM_DECLARE_EXTENDED_PHASE to retain the compatibility of assignment.

2.4.7 Base Protocol

The base protocol aims to maximal interoperability between components. It needs the usage of mentioned elements for different TLM classes/interfaces:

1. Core transport interface, DMI, debug transport interface,
2. Sockets (initiator/target),
3. Generic payload, and
4. TLM Phases.

Class tlm_base_protocol_types represents the base protocol. It is obligatory that all components should follow mentioned rules for maximal interoperability. In case new protocol is created, the coding styles and rules associated with the new protocol should follow the base one.

Phase Sequences

1. The protocol allows the use of blocking, non-blocking core transport interface, or even the both together.

Table 2.1 Permitted phase transitions [5]

Previous state	Calling path	Phase argument on call	Phase argument on return	Status on return	Response valid	End-of-life	Next state
//rsp	Forward	BEGIN_REQ	–	Accepted			req
//rsp	Forward	BEGIN_REQ	END _REQ	Updated			//req
//rsp	Forward	BEGIN_REQ	BEGIN_RESP	Updated	X		rsp
//rsp	Forward	BEGIN_REQ	–	Completed	X	X	//rsp
req	Backward	END_REQ	–	Accepted			//req
req	Backward	BEGIN_RESP	–	Accepted	X		rsp
req	Backward	BEGIN_RESP	END_RESP	Updated	X	X	//rsp
req	Backward	BEGIN_RESP	–	Completed	X	X	//rsp
//req	Backward	BEGIN_RESP	–	Accepted	X		rsp
//req	Backward	BEGIN_RESP	END_RESP	Updated	X	X	//rsp
//req	Backward	BEGIN_RESP	–	Completed	X	X	//rsp
rsp	Forward	END_RESP	–	Accepted	X	X	//rsp
rsp	Forward	END_RESP	–	Completed	X	X	//rsp

States Description: req, //req, rsp, //rsp stand for BEGIN_REQ, END_REQ, BEGIN_RESP, and END_RESP, respectively

2. There are constraints on non-blocking transport interface as it carries phase information. Thus, non-blocking interface method is more used for approximately timed coding style, while blocking transport method is more for loosely timed coding style.
3. The transitions of phases on a socket for transaction are
   BEGIN_REQ –> END_REQ –> BEGIN_RESP –> END_RESP.
4. For phases BEGIN_REQ and END_REQ, they are handled through initiator sockets, while BEGIN_RESP and END_RESP are handled through target sockets.

Permitted Phase Transitions

This is a summary for rules and the related phases transitions listed in Table 2.1.

2.4.8 TLM-2.0 Examples

Transition Sequence Flowchart Example

An example is illustrated in Fig. 2.18 [5] for non-blocking transport interface method calls for base protocol.

As mentioned before, the non-blocking interface breaks down the transaction into multiple phases. Each phase has its corresponding timing point, where each call/return can correspond to a phase transition. It supports both forward/backward paths for transactions between initiator and target. There are two declarations for

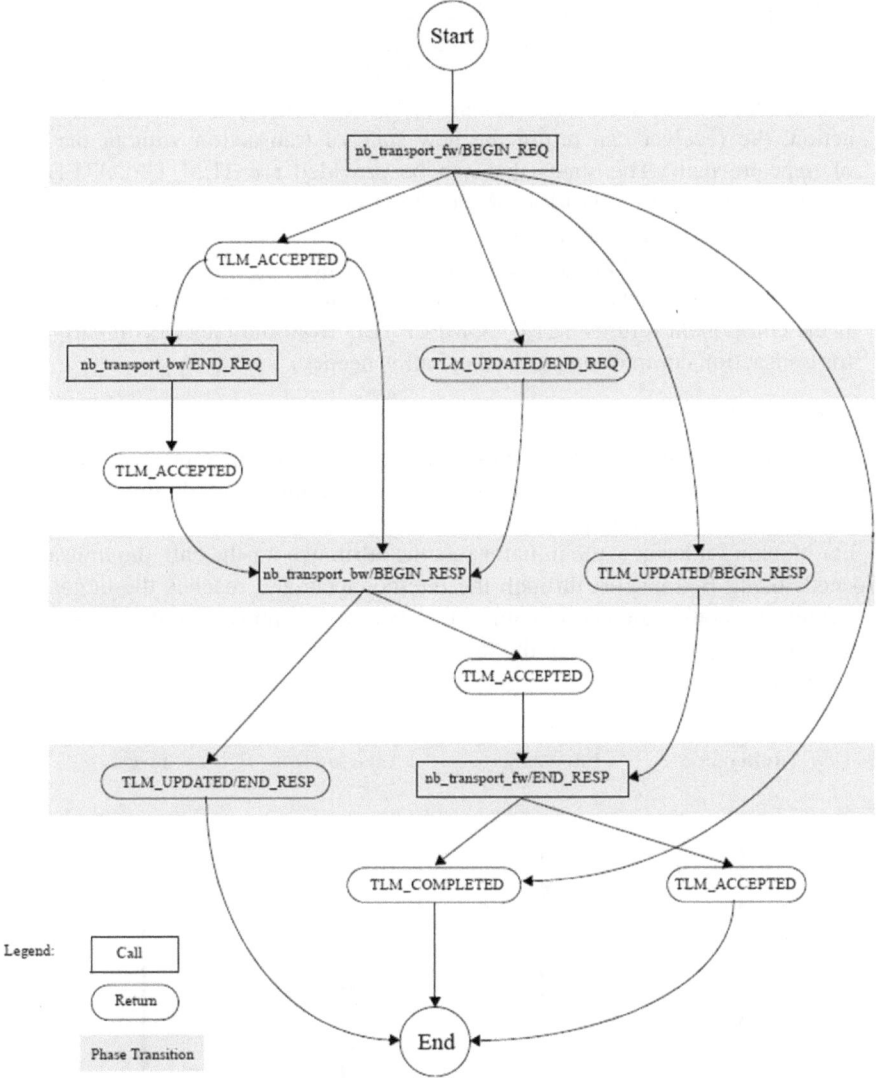

Fig. 2.18 Transition sequence example [5]

interface to support both directions. There are multiple potential paths for transaction communication between initiator and target according to the transaction setting. For approximately timed and base protocol, transaction is communicated back-and-forth couple of times between components.

1. If the transaction is not calculated immediately, return value should be TLM_ACCEPTED. Then, it would go through END_REQ using backward

path, where the return value should be TLM_ACCEPTED as well. After that, BEGIN_RESP is initiated at the target in backward path. Finally, the phase would be END_RESP at the initiator using TLM_ACCEPTED as return value.

2. On the other hand, if the receiver can expect immediately the behavior of transaction; the receiver can return the new state of transaction without the need of opposite path. The states that can be provided are TLM_UPDATED and TLM_COMPLETED. TLM_COMPLETED could be used at any stage if all transaction phases are already done.

3. Time delays could be annotated in calls to indicate when the transaction would be processed. The delays are annotated in calls in either direction.

4. If the component returns TLM_COMPLETED from the interface at early time for transaction completion, no further call is needed.

Causality Example

Examples for causality with blocking and non-blocking methods are shown in Fig. 2.19 [5] and Fig. 2.20 [5], respectively. The examples include initiator, target, and two interconnects' components.

For blocking interface, the initiator sets the attributes for the call, the transaction proceeds using b_transport through the interconnects and reaches the target. The interconnects modify the address attribute. Then, the response is initiated from target component, passing through the interconnects, and reaches the initiator as final destination.

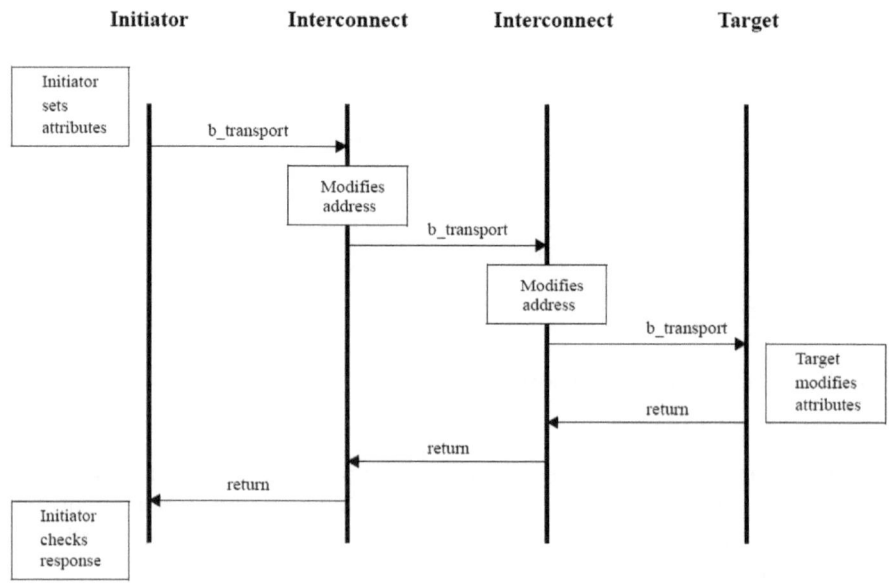

Fig. 2.19 Blocking interface method causality [5]

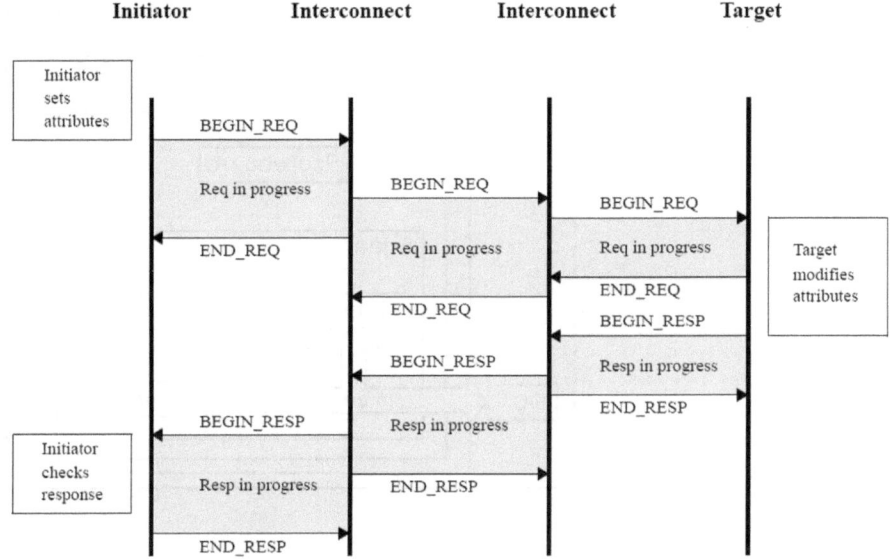

Fig. 2.20 Non-blocking interface method causality [5]

For non-blocking interface, the transaction goes through same components in phases. The request or response is processed between initiator/target (including interconnects).

2.5 Universal Verification Methodology

2.5.1 Introduction

The conventional manual RTL test benches are not reliable in verification of large SoCs. They are time-consuming and error-prone. Verification of those complex designs should be automated. Universal Verification Methodology (UVM) is an automatic functional verification environment for digital designs. It is based on design simulation, assertion, coverage, and random constraints for complete verification profile. A reference model is created for every specific verification plan. This reference model is highly abstracted and fast in simulation [14]. It should match the intended behavior of the original design. During simulation, the results of both the original design and reference models are evaluated and compared through a scoreboard, as illustrated in Fig. 2.21 [15].

A sequencer added by verification engineers creates stimulus test vectors according to the verification plan. Those test vectors are supplied to the driver. The driver module is connected to the design under test (DUT). The driver passes the test vec-

Fig. 2.21 UVM flow [15]

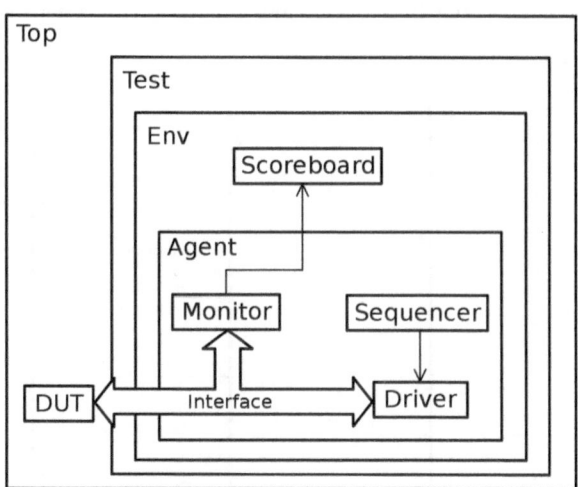

tors to this DUT. On the other hand, the monitor tracks the design output ports and delivers the results to the scoreboard. The scoreboard has a golden reference model for the design. Thus, the output of the design is compared with the golden reference. Hence, the design is verified for those particular test vectors. Several tests are created in UVM to have a complete verification of the design functionality for maximal achievable coverage.

2.5.2 Base Classes

The class library for UVM is available under a SystemVerilog package uvm_pkg. It contains different classes/definitions/utilities needed for reusable and salable verification environment. The building blocks for UVM environment are components and transactions. They are built using classes for the library:

1. uvm_object
2. uvm_component
3. uvm_transaction

uvm_object

The class represents the base for all UVM data and structure classes. It gives definitions for many methods needed for building the structure. Any component or transaction in UVM is derived from uvm_object. It determines the class-based core operations as create, copy, record, compare, and print. Also, it determines the information of instances as name, type, and seeding.

uvm_component

This is the base class for components in UVM. It provides the following interfaces:

1. Hierarchy:
   The user is able to traverse and search design hierarchy.
2. Phase:
   The test flow is phased which all components follow.
3. Reporting:
   All system messages, errors, and warning are processed through this interface.
4. Recording:
   All transaction processed by this component are recorded.
5. Factory:
   This is the interface to the factory for registration of components and objects.

uvm_transaction

This class is base for UVM transactions. It inherits uvm_object including all mentioned methods and it includes timing information and recording capability due to transient nature. Normal transaction is enabled/derived from uvm_transaction, while uvm_sequence_item is used for sequence-based ones.

2.5.3 UVM Factory

uvm_factory is used to build and store different UVM objects and components, such that only one instance is available at given simulation time. Upon request of new object/component from the factory, it determines the object type and creates an object from the same type.

2.5.3.1 Object/Component Proxies

The overhead of constructing an instance for every registered component and object is huge. Thus, the factory maintains wrappers/proxies; when a request happens for a new instance, the factory calls the proxy to create the needed instance.

The classes are derived from uvm_object (available under uvm_pkg: standard uvm package). They should have proxies declared as type_id. The type of proxy (proxy_type) could be

1. uvm_component_registry:
   Non-abstract derivative from uvm_component class. It has two forms for parameterized/non-parameterized usage.
2. uvm_abstract_component_registry:
   Abstract derivative from uvm_component class. It has two forms for parameterized/non-parameterized usage.

3. uvm_object_registry:
   Non-abstract derivative from uvm_object class. It has two forms for parameterized/non-parameterized usage.
4. uvm_abstract_object_registry:
   Abstract derivative from uvm_object class. It has two forms for parameterized/non-parameterized usage.

2.5.3.2 uvm_factory

This class is needed to create uvm objects and components. They are registered into the factory using proxies. uvm_factory is an abstract class in which many methods are purely virtual. It provides both name-based and type-based interfaces.

2.5.3.3 uvm_object_wrapper

This class enables the creation of object/component proxies using abstract interface. They are based on uvm_object and uvm_component and registered with uvm_factory.

2.5.3.4 uvm_phase

This needed for phasing the execution using automated mechanism for different design components. The class object has behavior, state, and schedule information.

UVM Common Phases

build_phase

This phase is needed to create and configure the design test bench.

connect_phase

This phase is needed to establish cross-component connections..

run_phase

This phase is needed to simulation run.

2.5.4 Predefined Classes

uvm_test

This is virtual class used to specify which test to be run in the design.

uvm_env

This is class used as hierarchical container of other components in order to build complete environment.

uvm_agent

This is class used to create user-defined agent.

uvm_monitor

This is class used to create UVM monitor component.

uvm_scoreboard

This is class used to create UVM scoreboard for needed comparisons.

uvm_driver

This is class used to create drivers which initiates requests for new transactions.

uvm_sequence_item

This class provides basic functionality for sequence items and sequences objects.

uvm_sequence

This class provides the interfaces needed to create streams of sequence items and sequences.

uvm_sequencer

This class is used as moderator for controlling transaction flow for different stimulus generators.

2.6 Design Synthesis

Synthesis is a stage in the design cycle where a translation of the high-level design model into Gate-Level (GL) cells takes place [16]. This high-level model is described in RTL, as it guarantees easier implementation, faster simulation, and reliable verification. This is hard to achieve on GL, due to the detailed gates' information. GL is created in either software netlist [17] or continues to hardware implementation.

After RTL design is verified with respect to the design specifications, the design is synthesized into netlist using different tools [18, 19]. The created netlist is technology-mapped information including timing information and internal gate parasitics.

Fig. 2.22 SAIF file syntax
format

```
(INSTANCE top
  (INSTANCE dut
    (NET
      (FRC1Data\[0\]
        (T0 999992)  (T1 8)  (TX 0)
        (TC 4)  (IG 0)
      )
```

2.7 Switching Activity Interchange Format

Every tool vendor and flow has a different database format for signals logging during simulations [20, 21]. However, simple standard formats are needed to ease the communication across tools and flows. There are several test file formats in order to store those needed signals. Value change dump (VCD) [22] and Switching Activity Interchange Format (SAIF) are examples of those formats. Both formats have timing information about every design signal during simulation. But they have a major difference; VCD has full information for the values at every instant for the whole simulation time, while SAIF has only the switching activity for the signal. The generated SAIF file is much smaller and easier in processing making it good candidate for power characterization.

For the example in Fig. 2.22, the "0" index of vector signal "FRC1Data" information is shown. This signal or net is under hierarchy of top-level "top" and instance "dut". The signal has five values of timing information as follows:

1. T0, T1, and TX represent total time spent at levels "0", "1", and "X", respectively.
2. TC is the toggle count between "0" and "1".
3. IG is the number of glitch transitions.

In TLPM, a SAIF file is used in the power characterization process.

2.8 Summary

Different fundamental concepts are introduced in this chapter. SystemC is an HDL based on C++ class library used mainly for system-level design. It provides different levels of design abstraction, which are needed for modeling of large SoCs, in order to have efficient and fast simulation. SystemC is used in Transaction-Level Modeling (TLM).

TLM is an abstract modeling of communication schemes among design modules. SoCs are built using TLM by creating modules and channels and integrating them together. Channels are modeled to handle the communication across modules including the core and peripherals. Functions are executed through calls from the interfaces.

Universal Verification Methodology (UVM) is an automatic functional verification environment for digital designs. It is based on design simulation, assertion, coverage, and random constraints for complete verification profile. A reference model is created for every specific verification plan. During simulation, the results of both the original design and reference models are evaluated and compared through a scoreboard.

Also, different design synthesis concepts/tools are discussed. SAIF file format and purposes are mentioned briefly at the end.

References

1. Accellera Systems Initiative. Universal Verification Methodology Language Reference Manual. http://www.accellera.org.
2. Accellera Systems Initiative. Universal Verification Methodology. http://www.accellera.org.
3. Intel Corporation. https://www.intel.com.
4. Jayadevappa, S., Shankar, R., & Mahgoub, I. (2004). A comparative study of modelling at different levels of abstraction in system on chip designs: A case study. In *Proceedings of IEEE Computer society Annual Symposium* (pp. 52–58).
5. Accellera Systems Initiative. Transaction Level Modeling. http://www.accellera.org.
6. Zhu, J. (2001). MetaRTL: Raising the abstraction level of RTL design. In *Proceedings of the Conference on Design, Automation and Test in Europe* (pp. 71–76).
7. Maillet-Contoz, L., & Ghenassia, F. (2005). Transaction level modeling-An abstraction beyond RTL. In *Transaction level modeling with systemC-TLM concepts and applications for embedded systems* (p. 23).
8. Accellera Systems Initiative. IEEE 1666–2011: SystemC Language Reference Manual. http://www.accellera.org.
9. Ashenden, P. J. (2010). *The designer's guide to VHDL* (Vol. 3). Burlington: Morgan Kaufmann.
10. Accellera Systems Initiative. IEEE 1800–2012: SystemVerilog Language Reference Manual. http://www.accellera.org.
11. Cesario, W. O., Lyonnard, D., Nicolescu, G., Paviot, Y., Yoo, S., Jerraya, A. A., et al. (2002). Multiprocessor SoC platforms: A component-based design approach. *IEEE Design and Test of Computers, 19*(6), 52–63.
12. ARM Ltd. ARM AMBA Advanced Microcontroller Bus Architecture. http://www.arm.com.
13. Burton, M., & Donlin, A. (2004). Transaction level modeling: Above RTL design and methodology. *Open systemC initiative TLM-working group.* http://www.systemc.org.
14. Salemi, R. (2013). The UVM primer: An introduction to the universal verification methodology.
15. Araujo, P. UVM Guide for Beginners. http://colorlesscube.com.
16. Borriello, G., & Detjens, E. (1988). High-level synthesis: Current status and future directions. In *Proceedings of ACM/IEEE Design Automation Conference* (pp. 477–482).
17. McFarland, M. C., Parker, A. C., & Camposano, R. (1990). The high-level synthesis of digital systems. *Proceedings of the IEEE, 78*(2), 301–318.
18. Cadence, Inc. SoC Encounter. https://www.cadence.com.
19. Synopsys, Inc. Design Compiler User Guide. http://www.synopsys.com.
20. Mentor Graphics, Inc. Wave Log Format(WLF) - ModelSim User Guide. https://www.mentor.com.
21. Synopsys, Inc. File System Data Base(FSDB) - VCS User Guide. http://www.synopsys.com.
22. IEEE: 1364–2001 - IEEE Standard Verilog Hardware Description Language. http://www.ieee.org.

Chapter 3
Power Modeling and Characterization

3.1 Introduction

Previous related works in power computation at different levels of design models are demonstrated in this chapter. It exposes the flow of previous methodologies in details including their pros and cons according to feasibility and cost of implementation. Also, potential contributions of the proposed TLPM methodology are presented.

Several methodologies have been developed to estimate the power using different flows at various abstraction levels of the design. Those methodologies have different approaches for power characterization of power parameters or the way of implementation. The methodologies are classified in the following sections according to the setup of work.

3.2 Power Modeling at Different Levels

3.2.1 System Level

Estimating power at the highest level of system abstraction has been addressed in [1]. Power characterization process in this work captures information for a particular core. This information is saved in a database called hierarchical transaction-level power (HTLP). This database is further used in logging the power information for the modules. This is performed through power mapping for the scenarios executed on system level using TLM. The overall flow for power estimation for this work is shown in Fig. 3.1 [1].

The gate-level netlist of the design including all parasitic information is supplied the characterization model. The test cases which are instructions/use cases for this design are used for evaluation of power model per each of them. The characterization is performed at low level of abstraction, building HTLP database. The database is mapped to the transaction representation for power model information. Thus, on

© Springer Nature Switzerland AG 2020
A. B. Darwish et al., *Transaction-Level Power Modeling*,
https://doi.org/10.1007/978-3-030-24827-7_3

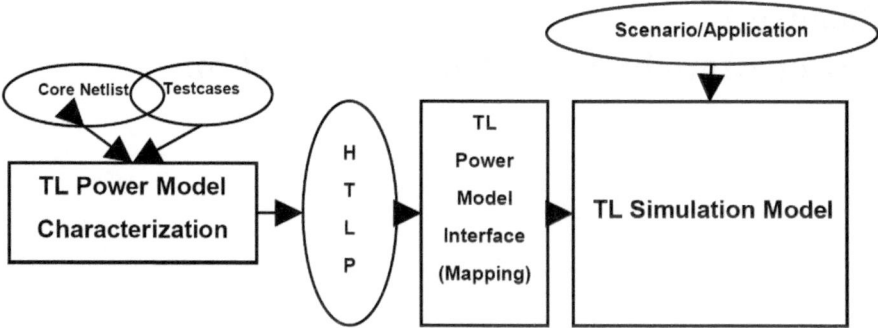

Fig. 3.1 HTLP power estimation flow [1]

applying the intended test cases of the design on TLM, the power information would be reported.

Power estimation is achieved by identifying the performed instructions according to the core functionality. Then, characterization is done for the power of each instruction using appropriate testing vectors. A macro-model is created for every instruction on system level. Thus, during TLM simulation, macro-models gather power updates per instruction [2, 3].

Another approach for power characterization using instructions is described in [4, 5]. In these works, the instructions are traced on different peripherals and after dependency checking between those instructions. Thus, characterization is performed according to the input data and power models are built at the system level.

In [6], another approach for power modeling on system level is adopted. The models are embedded inside behavioral Verilog modules. This approach is much slower in simulation than TLM [7, 8].

The mentioned approaches consider tracking the instructions for the core or peripherals on system level. Therefore, power estimation is not accurate, as it misses important contributors in the design with respect to power. Power management components are example for those elements. Thus, the output power numbers are not accurate as compared to the actual on-chip ones [9, 10].

However, the methodology presented in Chap. 4 is adopting a different approach. It implements the power models on the components themselves. This is achieved by characterizing each component separately. This includes the power numbers from all modules. Afterward, those components can be integrated for full SoC power estimation.

3.2.2 Component Level

In [11], power models are created in RTL for all components including simple functional blocks. The simulation time for this methodology is too long for two reasons.

The methodology works in RTL itself and creates/simulates power models of all components. Some components may have dummy functionality or negligible effect on power consumption. Alternatively, TLPM covers the main components in the design which have effective contribution in total power consumption.

3.3 Levels of Abstraction in Modeling

SoC designs are modeled at different levels of abstractions. This is necessary to accomplish different objectives across product design cycle [12, 13]. The timing information varies in the models according to cycle accuracy in the abstraction level [14]. The models are defined as cycle-accurate, approximately timed, or loosely timed.

The power kernel tool (PKtool) [15–17] adopts cycle-accurate models in TLM. An example of power module is illustrated in Fig. 3.2 [16]. The PKtool accepted the input signals of design, the changes in internal variables in the design from SC_MODULE, and predefined static information for power evaluation. The model checks all design registers at every clock cycle. Cycle-accurate models are time-consuming in simulation. Thus, it does not satisfy the required fast simulation of TLM [18, 19].

The research work in [20, 21] uses a different approach for TLM models of a design. It has a hybrid model in TLM. The methodology is implemented at different levels of abstraction. The methodology guarantees the control on the level of accuracy of each generated power model. It has two levels of granularity: coarse and fine. The coarse tuning is needed for the higher level of abstraction, in order to have reliable output results in reasonable amount of time. On the other hand, fine details are added to the lower level of abstraction. However, this hybrid model leads

Fig. 3.2 Power kernel tool [16]

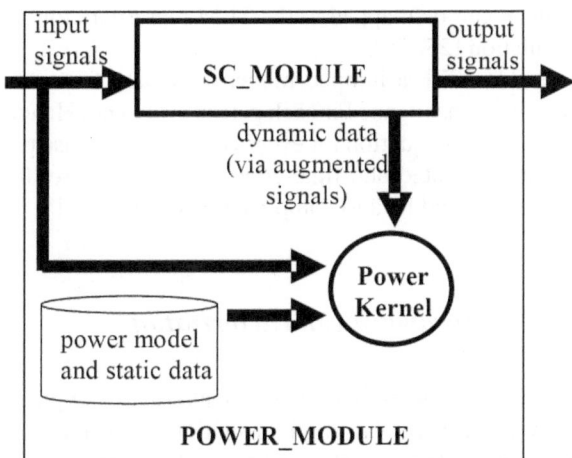

to several disadvantages. The methodology has sophisticated criteria to differentiate between coarse- and fine-tuning levels. This induces complexity in both power characterization process and TLM implementation of power models [22, 23].

The methodology in Chap. 4 adopts approximately timed power models. They are much faster in execution than [7, 8], and have reasonable accuracy in design modeling and simulation results as compared to [24].

In the next part, different approaches for power characterization are introduced, in order to create adequate power models with respect to accuracy and simulation time.

3.4 Power Characterization

Power characterization of the design is a fundamental stage in power evaluation process. This stage is needed in order to extract power parameters of the design elements at different operating conditions. The reliability of the extracted parameters values is crucial. However, the feasibility of implementation and time of extraction are significant factors in the process of power characterization. Various approaches for power characterization have been addressed in previous work as follows.

3.4.1 Gate-Level Netlist Characterization

Power characterization using extracted Gate-level netlist is a common practice. This approach is introduced in several publications such as [1, 16]. The work in [1] is performed on the core and peripherals. A gate-level netlist is generated as illustrated in Fig. 3.3 [1]. Then, placement and routing processes are executed and used to extract parasitic data. This data is manipulated with test vectors to build HTLP database. Thus, it can be mapped to the TLM representation in order to evaluate power upon simulation [25, 26].

In [16, 27], a lumped model is created for each type of gate. The equivalent parasitic capacitance for each gate is evaluated. Hence, the information is introduced in the energy equation for every component. Those processes of parasitics extraction are sophisticated and time-consuming. So, these different approaches are hard to implement and lead to complicated power model.

3.4.2 Hardware Characterization

In [28, 29], the authors consider alternative approach for power characterization. They use a hardware implementation of design in the laboratory where physical measurements are performed to read power parameters. An assembly program is

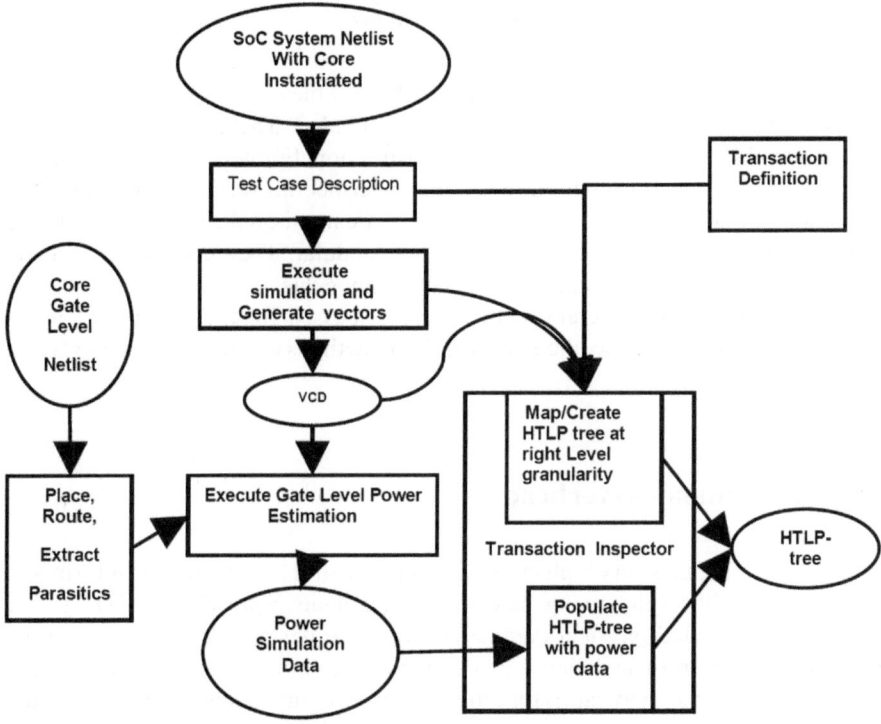

Fig. 3.3 HTLP power characterization flow [1]

supplied to the design to generate adequate stimulus. Furthermore, power models are
created according to the power parameters and integrated with the existing SystemC
modules. Upon simulation, energy consumption is tracked for every design module.
However, physical measurements need special setup and expertise, which decrease
feasibility of the methodology to be included in the industry.

3.4.3 Datasheet-Based Characterization

The publications in [30, 31] have a different approach in characterization. They
consider IP datasheet for the design for power evaluation. This datasheet is created
according to experiments performed after fabrication. IP datasheet is not adequate
solution especially in case of changing the used technology node. On the other hand,
power evaluation needs expert design and verification engineers to work on each
module [32].

3.4.4   Look-up Tables' Characterization

Instruction-based power modeling using look-up tables is adopted in [33]. First, the common instructions per peripheral are concluded. Then, GL simulations are executed to obtain the energy consumed per instruction. Furthermore, look-up tables, as in [34, 35], are constructed for instructions on peripherals versus the corresponding consumed power. At the model, the instructions are monitored and logged for power evaluation [36]. This kind of processes is not modular. They need to be executed from scratch based on the changes in design components.

In the next section, different methods are discussed for the addition of power models in TLM design. Also, the effects of those methods on the simulation overhead are presented.

3.5   Performance Overhead

Simulation overhead is very high in [37, 38]. In [37], the characterization information is associated to the behavior of hardware components represented in TLM. This information extracted physical parameters from the design. Upon simulation, the power consumption of associated components is computed. The characterization process is based on activity ratios for different design components. A generic function utilized for computation is function of voltage and frequency. The activity ratio represents an approximate ratio for gates activity between 0 and 1. Those activity ratios for all signals in design are evaluated by capturing data points all over the simulation. This adds extra overhead of up to 88% on simulation, which is very significant overload.

In [38], extending every SystemC module with add-on library leads to undesirable additional simulation time for power characterization. In this methodology, characterization process time depends on the number of instructions in the test.

3.6   Power Evaluation

Several approaches and flows have been developed in order to validate the design power specification. Those approaches are focusing either on verification of power management design techniques or evaluation of power consumption. For power management verification, several approaches are introduced. "Power Aware" [39, 40] is a verification technique for power management in RTL. It is responsible for verification of power switching, isolation checking, addition for proper retention cells, and level shifters.

For evaluation of power consumption, several methodologies are performed at different levels of abstraction [41] such as GL [42], transistor level [43, 44], and

Field-Programmable Gate Array (FPGA) [45, 46]. Accuracy of results is affected by the level of execution; however, the speed is the primary specification in this case [47].

Some approaches consider evaluation of the design at the hardware (HW) level. Experiments are held on real chip using devices in the laboratory for power characterization. On multiple iterations, the laboratory devices read power values and create a matrix for design components with the corresponding energy values [48, 49].

On the other hand, other approaches consider a pre-silicon FPGA option or emulators [50, 51] in power characterization [52, 53]. This approach is faster and cheaper than post-silicon option; however, it is less accurate.

Other approaches characterize the power in the software (SW) domain, as it is much cheaper in implementation than hardware. Also, power characterization on software level can address the issues at early stages in product development. SW post-silicon characterization can be modeled on different abstraction levels as instruction-based or algorithmic-based [54]. SW applications are applied to the real chip for power consumption calculation.

Furthermore, CAD tools are utilized for pre-silicon power characterization. Different flows and tools from different EDA companies are introduced in this area. One of the most well-known methodologies in power evaluation is power analysis flow [55] from Synopsys, Inc.

3.6.1 Power Analysis Flow

This flow is normally used for power calculation in RTL designs using the following steps:

1. **Generation of Signal Switching Activity**:
   RTL simulation is performed to obtain switching activity of every signal in the design. This activity is saved in a SAIF file mentioned in [2.7], which encloses toggling information of signals. This is achieved by several digital RTL simulators as in [56, 57].

2. **Design Synthesis**:
   RTL design is synthesized into netlist, which contains complete information of different GL components. Several tools are available for synthesis of the design and creation of GL netlist as in [58, 59].

3. **Power Calculation**:
   The SAIF file is attached to the synthesized netlist. Total power of the design is evaluated for the attached switching activity. Several tools do power evaluation after netlist attachment [55].

The flow is illustrated in Fig. 3.4.

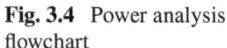

Fig. 3.4 Power analysis
flowchart

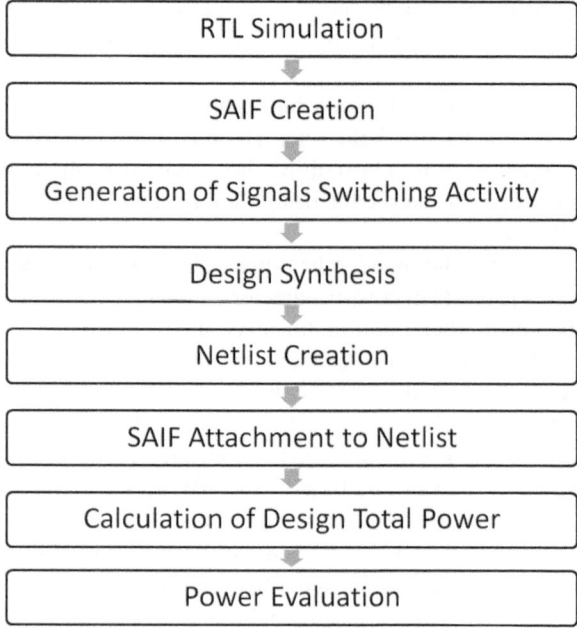

3.7   Summary

In this chapter, related work in power estimation in TLM is discussed in details.
Previous approaches are either too slow in simulation or too inaccurate in the characterization process especially for large SoC designs. Many approaches are based
on gate-level netlist or hardware itself for characterization. On the other hand, those
approaches suffer from performance overhead during TLM simulation due to sophisticated power models.

In this book, approximately timed modeling at high level of abstraction of the
design components is adopted to achieve fast simulation with high precision. Approximately timed model contains implementation details of the design and the mapping
relations between processes without the need of cycle-accurate RTL. For power
characterization, a lumped parameter is extracted for the signals of the design. The
extraction algorithm of the power parameters is simple and fast using power analysis flow. On the other hand, the simulation overhead of power characterization is
minimal.

A methodology for estimating power dissipation on TLM (TLPM) is developed.
Having a power model for different small design components allows estimating
power of large designs of SoC with reasonable accuracy.

References

1. Narayanan, V., & Dhanwada, N. (2005). A power estimation methodology for systemC transaction level models. In *Proceedings of IEEE/ACM/IFIP International Conference on Hardware/Software Codesign and System Synthesis* (pp. 142–147).
2. Bona, A., Zaccaria, V., & Zafalon, R. (2004). System level power modeling and simulation of high-end industrial network-on-chip. *Ultra low-power electronics and design* (pp. 233–254). US: Springer.
3. Brooks, D., Tiwari, V., & Martonosi, M. (2000). *Wattch: A framework for architectural-level power analysis and optimizations* (Vol. 28.2, pp. 83–94).
4. Qu, G., Kawabe, N., Usami, K., & Potkonjak, M. (2000). Function-level power estimation methodology for microprocessors. In *Proceedings of Annual Design Automation Conference* (pp. 810–813).
5. Bergamaschi, R. A., & Jiang, Y. W. (2003). State-based power analysis for systems-on-chip. In *Proceedings of annual Design Automation Conference* (pp. 638–641).
6. Benini, L., Hodgson, R., & Siegel, P. (1998). System-level power estimation and optimization. In *Proceedings of the International Symposium on Low power electronics and design* (pp. 173–178).
7. Cai, L., & Gajski, D. (2003). Transaction level modeling: An overview. In *Proceedings of IEEE/ACM/IFIP International Conference on Hardware/Software Codesign and System Synthesis* (pp. 19–24).
8. Keutzer, K., Newton, A. R., Rabaey, J. M., & Sangiovanni-Vincentelli, A. (2000). System-level design: orthogonalization of concerns and platform-based design. *IEEE transactions on computer-aided design of integrated circuits and systems, 19*(12), 1523–1543. December.
9. Jayadevappa, S., Shankar, R., & Mahgoub, I. (2004). A comparative study of modelling at different levels of abstraction in system on chip designs: a case study. In *Proceedings of IEEE Computer society Annual Symposium* (pp. 52–58).
10. Lajolo, M., Raghunathan, A., & Dey, S. (2000). Efficient power co-estimation techniques for system-on-chip design. In *Proceedings of the Conference on Design, Automation and Test in Europe* (pp. 27–34).
11. Raghunathan, A., Dey, S., & Jha, N. K. (1997). Register-transfer level estimation techniques for switching activity and power consumption. In *Proceedings of the 1996 IEEE/ACM International Conference on Computer-Aided Design* (pp. 158–165).
12. Ye, W., Vijaykrishnan, N., Kandemir, M., & Irwin, M. J. (2000). The design and use of simplepower: a cycle-accurate energy estimation tool. In *Proceedings of Design Automation Conference* (pp. 340–345).
13. Chen, R. Y., Irwin, M. J., & Bajwa, R. S. (2001). Architecture-level power estimation and design experiments. *Transactions on Design Automation of Electronic Systems (TODAES), 6*(1), 50–66. January.
14. Simunic, T., Benini, L., & De Micheli, G. (1999). Cycle-accurate simulation of energy consumption in embedded systems. In *Proceedings of Design Automation Conference* (pp. 867–872).
15. Universita Politecnica della Marche. PKtool Power Kernel Tool Documentation. http://pktool.sourceforge.net.
16. Vece, G. B., & Conti, M. (2009). Power Estimation in Embedded Systems within a SystemC-based Design Context: the PKtool Environment. In *Intelligent solutions in Embedded Systems* (pp. 179–184).
17. Greaves, D., & Yasin, M. (2014). TLM POWER3: Power estimation methodology for systemC TLM 2.0. In *Models, Methods, and Tools for ComplexChip Design* (pp. 53–68). Springer International Publishing.
18. Clouard, A., Jain, K., Ghenassia, F., Maillet-Contoz, L., & Strassen, J. P. (2003). Using transactional level models in a SoC design flow. *SystemC* (pp. 29–63). US: Springer.

19. Donlin, A. (2004). Transaction level modeling: flows and use models. In *Proceedings of the 2nd IEEE/ACM/IFIP International Conference on Hardware/Software Codesign and System Synthesis* (pp. 75–80).
20. Atitallah, R. B., Niar, S., Greiner, A., Meftali, S., & Dekeyser, J. L. (2006). Estimating energy consumption for an MPSoC architectural exploration. In *International Conference on Architecture of Computing Systems* (pp. 298–310). Berlin, Heidelberg: Springer.
21. Atitallah, R., Niar, S., & Dekeyser, J. (2007). MPSoC power estimation framework at transaction level modeling. In *Proceedings of the Microelectronics International Conference*.
22. Givargis, T., Vahid, F., & Henkel, J. (2000). A hybrid approach for core-based system-level power modeling. In *Proceedings of Asia and South Pacific Design Automation Conference* (pp. 141–146).
23. Lee, I., Kim, H., Yang, P., Yoo, S., Chung, E. Y., Choi, K. M., Kong, J. T., & Eo, S. K. (2006). PowerViP: Soc power estimation framework at transaction level. In *Proceedings of Asia and South Pacific Design Automation Conference* (pp. 551–558).
24. Pazos, N., Brunnbauer, W., Foag, J., & Wild, T. (2003). System level performance estimation of multi-processing, multi-threading SoC architectures for networking applications. *SystemC* (pp. 157–190). US: Springer.
25. Bergamaschi, R. A., Shin, Y., Dhanwada, N., Bhattacharya, S., Dougherty, W. E., Nair, I., Darringer, J., & Paliwal, S. (2003). SEAS: A system for early analysis of SoCs. In *Proceedings of the 1st IEEE/ACM/IFIP International Conference on Hardware/Software Codesign and System Synthesis* (pp. 150–155).
26. Dekeyser, J. L., Niar, S., Meftali, S., & Atitallah, R. B. (2007). An MPSoC performance estimation framework using transaction level modeling. In *IEEE International Conference of Embedded and Real-Time Computing Systems and Applications* (pp. 525–533).
27. Najm, F. N. (1993). Transition density: A new measure of activity in digital circuits. *IEEE Transactions on Computer-Aided Design of Integrated Circuits and Systems, 12*(2), 310–323. February.
28. Tiwari, V., Malik, S., Wolfe, A., & Lee, M. T. C. (1996). Instruction level power analysis and optimization of software. In *Technologies for wireless computing* (pp. 139–154). US: Springer.
29. Varma, A., Debes, E., Kozintsev, I., Klein, P., & Jacob, B. (2008). Accurate and fast system-level power modeling: An XScale-based case study. *ACM Transactions on Embedded Computing Systems (TECS), 7*(3), 25. April.
30. Bansal, N., Lahiri, K., Raghunathan, A., & Chakradhar, S. T. (2005). Power monitors: A framework for system-level power estimation using heterogeneous power models. In *International Conference on VLSI Design* (pp. 579–585).
31. Lebreton, H., & Vivet, P. (2008). Power modeling in SystemC at transaction level, application to a DVFS architecture. In *Symposium on VLSI, IEEE Computer Society Annual* (pp. 463–466).
32. Lajolo, M., Raghunathan, A., Dey, S., & Lavagno, L. (2002). Cosimulation-based power estimation for system-on-chip design. *IEEE Transactions on Very Large Scale Integration (VLSI) Systems, 10*(3), 253–266.
33. Givargis, T., Vahid, F., & Henkel, J. (2002). Instruction-based system-level power evaluation of system-on-a-chip peripheral cores. *IEEE Transactions on Very Large Scale Integration (VLSI) Systems, 10*(6), 856–863.
34. Gupta, S., & Najm, F. (2000). Power macromodeling for high level power estimation, 34rd ACM. In *IEEE Transactions on Very Large Scale Integration (VLSI) Systems, 8*(1), 18–29.
35. Barocci, M., Benini, L., Bogliolo, A., Ricco, B., & De Micheli, G. (1999). Lookup table power macro-models for behavioral library components. In *Proceedings of IEEE Alessandro Volta Memorial Workshop on Low-Power Design,* (pp. 173–181).
36. Li, Y., & Henkel, J. (1998). A framework for estimation and minimizing energy dissipation of embedded HW/SW systems. In *Proceedings of annual Design Automation Conference,* (pp. 188–193).
37. Helmstetter, C., Bouhadiba, T., Moy, M., & Maraninchi, F. (2006). Fast and modular transaction-level-modeling. *IEEE Transactions on VLSI Systems,* 501–513

38. Bouhadiba, T., Moy, M., & Maraninchi, F. (2013). System-level modeling of energy in TLM for early validation of power and thermal management. In *Proceedings of the Conference on Design, Automation and Test in Europe*, (pp. 1609–1614).
39. Mentor Graphics, Inc. Low Power Design. https://www.mentor.com.
40. IEEE. 1801-2015 - IEEE Standard for Design and Verification of Low-Power, Energy-Aware Electronic Systems. http://www.ieee.org.
41. Zhong, L., Ravi, S., Raghunathan, A., & Jha, N. K. (2004). Power estimation for cycle-accurate functional descriptions of hardware. In *Proceedings of the 2004 IEEE International conference on Computer-aided design*, (pp. 668-675).
42. Saleh, R. A., & Newton, A. R. (1990). Gate-level simulation. *Mixed-Mode Simulation* (pp. 101–132). US: Springer.
43. Wang, X., & Porod, W. (2000). Single-electron transistor analyticI Vmodel for SPICE simulations. *Superlattices and Microstructures, 28*(5), 345–349.
44. Cao, Y., Sato, T., Orshansky, M., Sylvester, D., & Hu, C. (2000). New paradigm of predictive MOSFET and interconnect modeling for early circuit simulation. In *Proceedings of the IEEE Custom Integrated Circuits Conference*, (pp. 201–204).
45. Brown, S. (1996). FPGA architectural research: A survey. *IEEE Design and Test of Computers, 13*, 9–15. December.
46. Xilinx, Inc. Virtex FPGA series. https://www.xilinx.com.
47. Nemani, M., & Najm, F. N. (1996). Towards a high-level power estimation capability. *IEEE Transactions on Computer-Aided Design of Integrated Circuits and Systems, 15*(6), 588–598.
48. Reda, S. (2011). Thermal and power characterization of real computing devices. *IEEE Journal on Emerging and Selected Topics in Circuits and Systems, 1*(2), 76–87.
49. McCullough, J. C., Agarwal, Y., Chandrashekar, J., Kuppuswamy, S., Snoeren, A. C., & Gupta, R. K. (2011). Evaluating the effectiveness of model-based power characterization. In *USENIX Annual Technical Conference*, (vol. 20).
50. Mentor Graphics, Inc. Veloce Emulation Platform. https://www.mentor.com.
51. Coburn, J., Ravi, S., & Raghunathan, A. (2005). Hardware accelerated power estimation. *Proceedings of the conference on Design, Automation and Test in Europe, 1*, 528–529. March.
52. Sunwoo, D., Wu, G. Y., Patil, N. A., & Chiou, D. (2010). PrEsto: An FPGA-accelerated power estimation methodology for complex systems. In *Field Programmable Logic and Applications (FPL) International Conference*, (pp. 310–317).
53. Degalahal, V., & Tuan, T. (2005). Methodology for high level estimation of FPGA power consumption. In *Proceedings of the 2005 Asia and South Pacific Design Automation Conference*, (pp. 657–660).
54. Reda, S., & Nowroz, A. N. (2012). Power modeling and characterization of computing devices. *Foundations and Trends in Electronic Design Automation, 6*(2), 121–216. May.
55. Synopsys, Inc. Power Compiler User Guide. http://www.synopsys.com.
56. Mentor Graphics, Inc. QuestaSim User Guide. https://www.mentor.com.
57. Cadence, Inc. Incisive Unified Simulator. https://www.cadence.com.
58. Cadence, Inc. SoC Encounter. https://www.cadence.com.
59. Synopsys, Inc. Design Compiler User Guide. http://www.synopsys.com.

Part II
Transaction-Level Power Modeling

Chapter 4
Transaction-Level Power Modeling Methodology

4.1 Introduction

In this chapter, Transaction-Level Power Modeling (TLPM) methodology is proposed for dynamic power estimation using TLM. The TLPM methodology is qualified to be easily implemented using industrial flows since it utilizes commercial tools.

First, an overview of the proposed flow is introduced. TLPM is performed on two stages: characterization and implementation. The characterization stage includes the extraction of power parameters. Then, implementation and simulation of power model are demonstrated.

4.2 TLPM Flow

4.2.1 Overview

TLPM or Transaction-Level Power Modeling methodology is proposed in this book for dynamic power estimation of SoC designs using TLM. The methodology exploits the existing tools for RTL simulation, debugging, design synthesis, and SystemC prototyping to provide fast and accurate power estimation results. The complete flow is illustrated in Fig. 4.1.

The methodology is performed in two stages: power characterization and TLM implementation. Power characterization is the extraction of power parameters of the RTL design. Power parameter defines the exerted energy by every design element. This is achieved by applying switching activities on the synthesized design and processing input data. Correlation between the signals with respect to power consumption is captured. A database called the correlation matrix is created to log this information. This database has all RTL signals with the corresponding exerted energy

© Springer Nature Switzerland AG 2020
A. B. Darwish et al., *Transaction-Level Power Modeling*,
https://doi.org/10.1007/978-3-030-24827-7_4

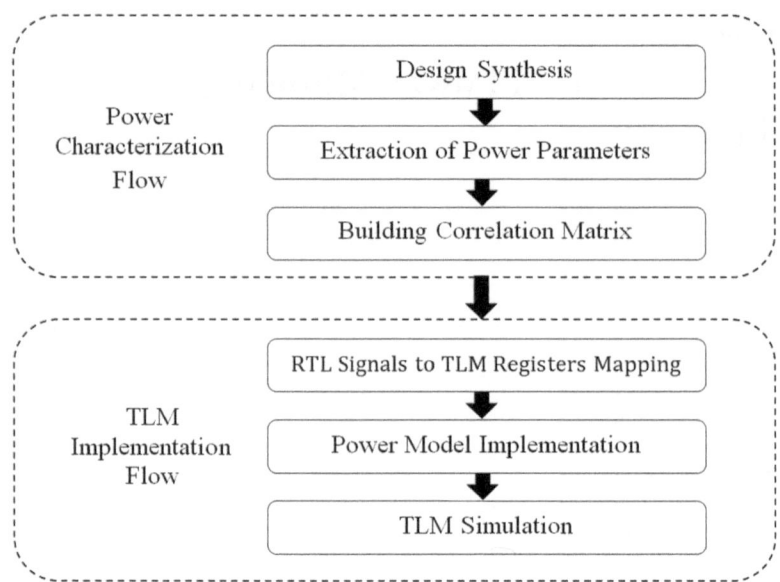

Fig. 4.1   Transaction-level power modeling flow

and their cross correlation. The correlation matrix is supplied to the implementation phase based on TLM.

TLPM implementation flow starts with the mapping of RTL signals onto registers in the TLM model. The information of the correlation matrix is applied to the TLM registers. Thus, every register in the TLM has its own power parameter. This parameter represents the exerted energy by a particular register in a certain operation. Power models are built for all registers exploiting the energy information from the correlation matrix. Finally, during simulation of the TLM model enclosing the power models, dynamic power consumption is evaluated.

4.2.2   Power Estimation Using TLPM

For large SoCs, simulation of the GL netlist to perform power evaluation is not feasible. A fast TLM simulation is proposed to achieve this target. Bottom-up approach is adopted in this book to perform power estimation of large SoCs described in TLM. This process is illustrated in Fig. 4.2.

A power model based on TLM is created for different units/components. Those components are integrated to create the SoC model. This allows estimating power of large designs in a reasonable simulation time. Building power model in TLM needs power characterization in order to calculate the exerted energy by different elements in the design.

Fig. 4.2 Power estimation for large SoCs using TLPM

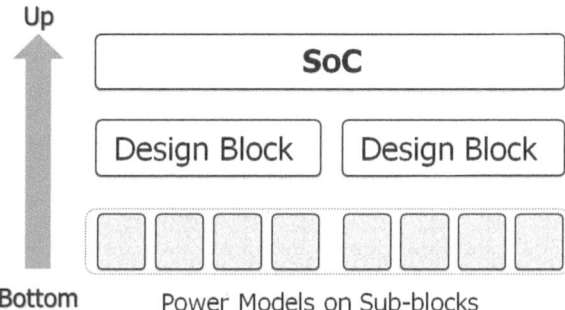

Upon simulation of the components, the models gather energy consumption over-time for the whole SoC. In this book, the implementation and simulation of power model for a single component are discussed.

4.3 Power Characterization

In the TLPM methodology, power characterization represents the main building block for power model creation. The objective of the characterization process is to identify the main contributors of power consumption at different loads of the design. This is achieved by the extraction of power parameters for different design components or signals. Power characterization process on TLPM receives RTL design as an input. After design processing, a correlation matrix database is created at the end of the process. Power characterization process is summarized in Fig. 4.3.

Power analysis flow, refer to Sect. 3.6.1, is utilized to track the power consumption of every signal to deduce the power parameters. After performing the dependency checking between the signals, a correlation matrix is created.

4.3.1 TLPM Power Characterization Flow

Power analysis flow is adopted to perform power characterization. Power analysis flow evaluates power by attaching RTL signals switching activity from a SAIF file to the GL netlist. Power characterization is achieved by calculating exerted energy by every design element, which has its own contribution to the total power. Using the superposition concept, the total power is divided among its contributors of the design elements. This is performed by splitting down the design into segments. Those segments represent linear equations between elements. During the implementation phase, those elements are mapped to the corresponding registers in the TLM model. Then, a comprehensive estimation is achieved for the total power. Characterization process is illustrated in Fig. 4.4.

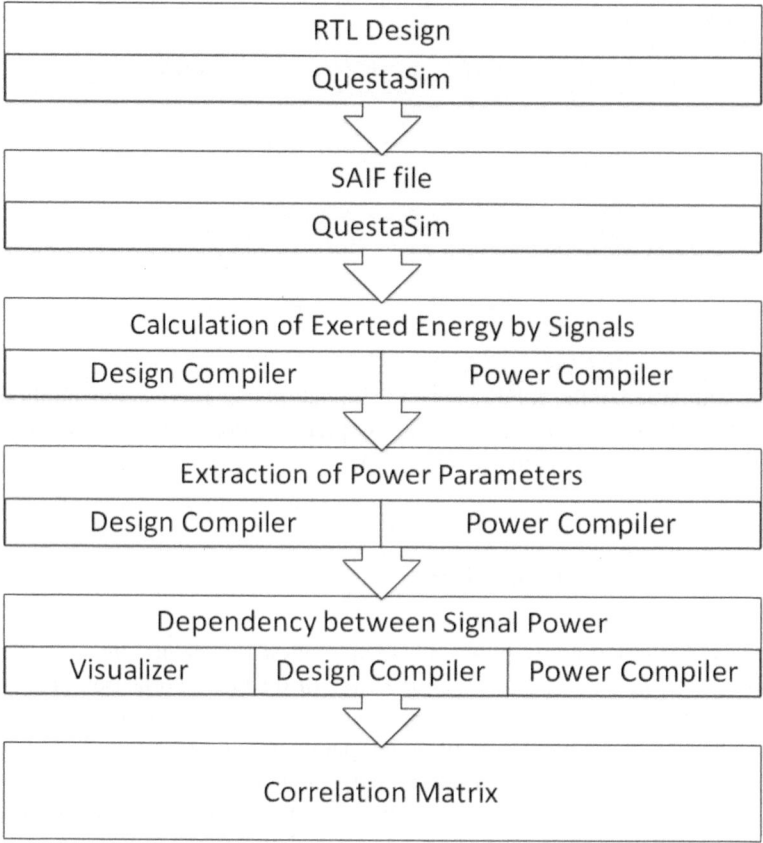

Fig. 4.3 TLPM power characterization process and EDA tools

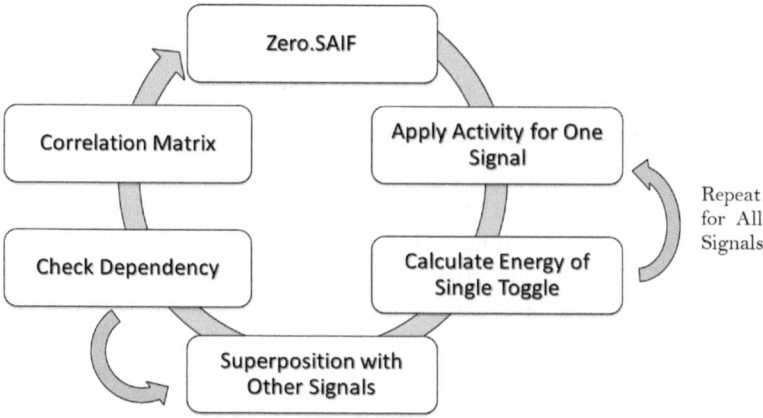

Fig. 4.4 Characterization process

In order to achieve power characterization, the following steps are performed:

1. Create a Zero.SAIF file. It is the normal SAIF file of the design with zero activity information for all RTL signals.
2. Adopt superposition theory for the circuit elements. Enable the activity of only one element at a time, to evaluate contribution of each element in total power.
3. Apply switching activity on a single element in a Zero.SAIF file. Power information of every signal is evaluated in the RTL model, where the energy of a single toggle count is determined. This step is repeated for all signals.
4. In order to overcome dependencies among signals, signal activity is added incrementally into the Zero.SAIF file. The dependency scheme among RTL signals is completed upon performing this step for all signals.
5. The dependency scheme is divided into three correlation factors:

    (a) Zero correlation factor: The signal has its own distinct contribution in total dynamic power, and no other signal shares its contribution in total power value. Clock tree signals are good examples of those.
    (b) One correlation factor (total dependency): In this case, signals have the same contribution with respect to power. They are grouped in one bundle. Signals on the same datapath match this category.
    (c) Nonzero, non-one correlation factor: In this case, signals have some effect on each other. Superposition cannot be used, and therefore conditional correlation is applied. Multiplexer outputs are an example for this case.

6. The dependency scheme between signals constitutes the "Correlation Matrix" of the design.
7. Iterate on these steps till covering all RTL signals.

An example is illustrated in Algorithm 4.1.

Algorithm 4.1 Conditional Correlation Example

Signals:

1. 'In_1' , 'In_2' $\Rightarrow$ contributors/inputs to the design
2. 'C' $\Rightarrow$ control signal in the design
3. 'Out' $\Rightarrow$ the affected signal

Energy contribution of In_1 and In_2 are 100, 200 respectively, and it is negligible for Control.

If $(value(C) = 0)$ **then**

$\Rightarrow$ Both 'In_1' and 'In_2' are both affecting 'Out' signal power/energy.
$\Rightarrow$ Energy(Out) = Energy(In_1) + Energy(In_2) = 100 + 200 = 300 energy unit.

else if $(value(C) = 1)$ **then**

$\Rightarrow$ In_1 only is affecting Out.
$\Rightarrow$ Energy(Out) = Energy(In_1) = 100 energy unit.

end if

According to the example, the energy exerted on a signal is subject to the energy exerted on other signals. Those signals are considered contributors for the energy equation for the signal of interest. The exerted energy on this signal could vary according to the input datapath. The datapath changes over simulation according to the setup/configuration at the time step based on control signal value. Thus, the value of energy exerted on signal is function in the exerted energy on input signals and the corresponding value of control signal. At the end, the needed information is exerted energy of a given signal at different configurations.

The correlation matrix database is built after completion of the characterization process. It contains all information of RTL signals: name, hierarchy, energy exerted for a single toggle count, and its correlation with other signals.

The process of power characterization is based on Perl [1] scripts. This Perl environment is used to process the SAIF file, check the dependency between the signals, and log the information into the correlation matrix. A snippet of the developed Perl script is listed in Fig. 4.5.

This code is responsible for applying activity of certain signals and merge it with a Zero.SAIF file. Thus, user can interpret the exerted power for certain signals on applying the generated file to power compiler for the experiments.

4.3.2   Correlation Matrix Example

Considering a simple example for RTL design as shown in Fig. 4.6. The design has two input ports from the outside world "input_1" and "input_2". They are not point of interest; however, they are affecting the internal signals in the design. The signals that are affected directly are "a" which is located under "top" and "b" which is located under "top/dut".

On performing power characterization process on this design, signal "a" has no dependency on any other signal for the value of exerted energy. The exerted energy value per toggle count is 10n. This value is the same through all configuration changes. While for signal "b" located under "top/dut", it is subject to configuration changes, because an extra load "c" is added to the datapath. The signal depends on a "Control" signal such that the energy holds values 30 or 40n based on "Control" value.

An example for correlation matrix table is listed in 4.1 for signal and the corresponding exerted energy according to certain setup.

4.3.3   EDA Tools for Power Characterization

Different tools are exploited in the TLPM power characterization flow. An overview is presented for those tools with respect to the common flow and usage.

```perl
my $file_in = in.saif;
my $file_out = out.saif;

open (FROM_FILE, config_file);
my @config_array = <FROM_FILE>;
close (FROM_FILE);

my $flag=0;
my $line = 1;

while(1) {
    my $line_content = `head \-${line} 0\.saif \|tail \-1`;
    chomp($line_content);
    $line_3 = $line + 3;
    my $flag=0;
    foreach my $element (@config_array) {
        chomp ($element);
        if ($line_content =~ m/\Q$element\E/) {
            $flag=1;
        }
    }
    if ($flag==1) {
        `sed -n '${line},${line_3}p;${line_3}q' in.saif >> out.saif`;
        $line = $line_3;
    } else {
        `sed -n '${line}p' 0.saif >> out.saif`;
    }
    $line++;
    if ($line==4587) {
        last;
    }
}
```

Fig. 4.5 Perl script snippet for SAIF file processing

Fig. 4.6 Correlation
example—RTL design

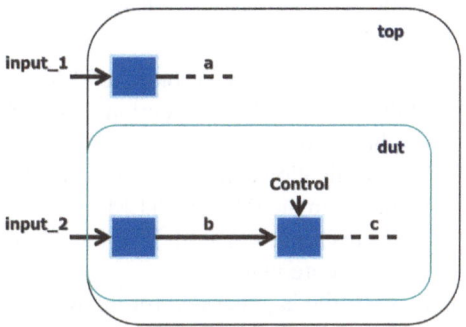

Table 4.1 Correlation matrix example

Signal name	Signal path	Energy value per toggle count	Condition
a	top	10 n	1
b	top/dut	30 n	Control == 0
b	top/dut	40 n	Control == 1
Control	top/dut	Negligible	NA
c	top/dut	20 n	1

4.3.3.1   QuestaSim ®

Overview

QuestaSim [2] is a Mentor (Siemens) tool for simulation and verification of digital designs. It has debugging and functional coverage capabilities for different HDLs such as SystemVerilog, VHDL and SystemC, also standards such as Unified Power Format (UPF) [3] and UVM [4]. Also, it supports different levels of abstraction as GL, RTL, and TLM.

Standard Flow

The design is supplied to the tool as HDL files. The flow is described in Fig. 4.7 as follows:

1. Compilation to parse the HDL files, checking for syntax, and building the design units.
2. Optimization of the design hierarchy for performance maximization along with setting of appropriate visibility for design debugging.
3. Launching the simulator and logging the needed signals.
4. Running the simulation to execute the design and log the signals values.

TLPM Usage

In the presented approach, a test bench for RTL is written in UVM format in order to serve different purposes:

1. UVM makes it easy to compare TLM and RTL results with respect to the accuracy of the model and the execution time. The RTL model represents golden reference for the TLM one.
2. The verification capability is used to check if there is any difference in functionality between the RTL and TLM models. This is performed by running class-based UVM test bench on both RTL and TLM, and checking the values of their signals using a scoreboard.
3. A SAIF file is generated for the RTL design. In standard usage, the file encloses RTL signals with the corresponding switching activity. In TLPM methodology, the SAIF file is processed for power characterization over several iterations, the exerted energy for every signal is deduced. Those energy values are logged in the correlation matrix and used in TLPM implementation process.

Fig. 4.7 QuestaSim flow

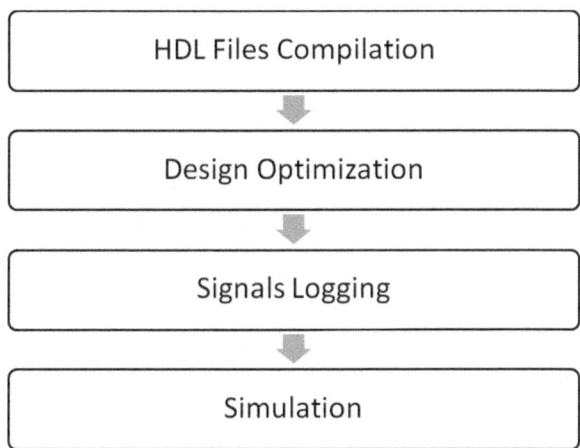

4.3.3.2   Design Compiler ®

Overview

Design compiler is a Synopsys Inc. tool for RTL synthesis based on a specific technology node. It is characterized by comprehensive optimization for power, area and timing for the design. Also, it has the capability for floor-plan exploration. The output of the tool is correlated within 10% of the real physical implementation, making it a good candidate for simple design synthesis solution in the TLPM methodology.

Standard Flow

Synthesizable RTL files and technology library are the inputs to the tool. The technology library is needed to map the RTL components into standard cells/gates according to the specified technology node and process conditions [5]. The flow is described in Fig. 4.8 as follows:

1. Analyze the design files in order to check for syntax and build the design units.
2. Elaborate the main design units into the tool kernel.
3. Clock tree annotation in the design.
4. Set the constraints for design optimization with respect to gates fan out, loads, power, area, and leakage.
5. Compile the design for mapping using the specified technology library.
6. Netlist file creation.

TLPM Usage

TLPM methodology uses design compiler for RTL design synthesis within the adapted power using the power compiler tool.

Fig. 4.8 Design compiler
flow

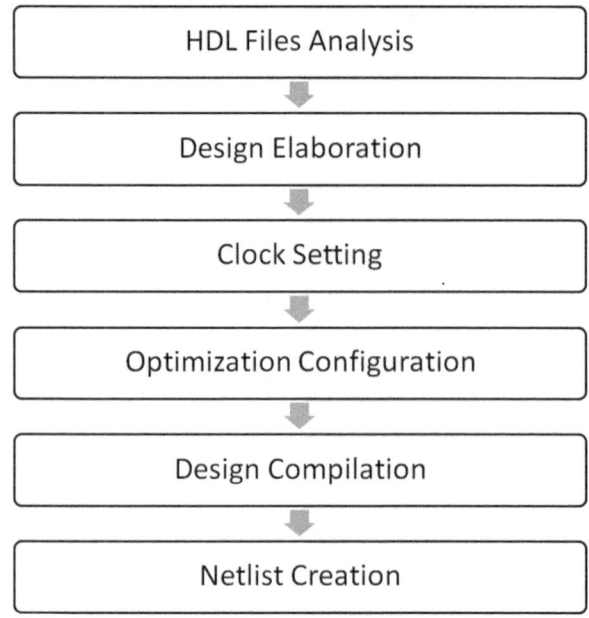

4.3.3.3 Power Compiler ®

Overview

Power compiler is a Synopsys Inc. tool for RTL and GL power consumption optimization and evaluation. It facilitates dynamic and leakage power optimization through various approaches such as clock gating and low-power techniques as power gating and retention cells. An example of output report is shown in Fig. 4.9.

Power Report Description

The report from power compiler has three sections:

1. Operating environment and units information:

    (a) Supply voltage used in evaluation.
    (b) Units reported for voltage, capacitance, time, and power (dynamic/leakage).

2. Summary for power results and ratios:

    (a) Dynamic power numbers for cells and nets.
    (b) Leakage power in cells.

3. Detailed information for power consumption contributors. It has matrix for power group contributor and type of power consumed by this group.

```
Global Operating Voltage = 1.2
Power-specific unit information :
    Voltage Units = 1V
    Capacitance Units = 1.000000ff
    Time Units = 1ns
    Dynamic Power Units = 1uW      (derived from V,C,T units)
    Leakage Power Units = 1pW


  Cell Internal Power   =   74.4402 uW    (70%)
  Net Switching Power   =   32.6640 uW    (30%)
                            ---------
Total Dynamic Power     =  107.1042 uW   (100%)

Cell Leakage Power      =   20.5260 uW
```

Power Group	Internal Power	Switching Power	Leakage Power	Total Power	(	%	)
io_pad	0.0000	0.0000	0.0000	0.0000	(	0.00%	)
memory	0.0000	0.0000	0.0000	0.0000	(	0.00%	)
black_box	0.0000	0.0000	0.0000	0.0000	(	0.00%	)
clock_network	0.0000	0.0000	0.0000	0.0000	(	0.00%	)
register	4.7950	16.5884	9.8097e+06	31.1931	(	24.44%	)
sequential	0.0000	0.0000	0.0000	0.0000	(	0.00%	)
combinational	69.6452	16.0757	1.0716e+07	96.4372	(	75.56%	)
Total	74.4402 uW	32.6640 uW	2.0526e+07 pW	127.6302 uW			

Fig. 4.9 Power report

(a) Power consumption groups:
 i. Pads.
 ii. Memories.
 iii. Black box (i.e., secured IPs).
 iv. Clock network.
 v. Design registers.
 vi. Sequential logic blocks.
 vii. Combinational logic blocks.
(b) Power consumption type:
 i. Cell internal dynamic power.
 ii. Switching dynamic power on nets.
 iii. Leakage power of cells.

According to the report, most of dynamic power is consumed in cells of the combinational logic blocks.

Common Flow

A synthesized design or netlist is supplied to the power compiler along with the SAIF file, which encloses the toggling information of the RTL signals. Then, the tool reports the power consumption. The flow is as follows:

1. Read the synthesized design or netlist.
2. Attach the SAIF file containing the toggling activity of the RTL design signals.
3. Power consumption is reported according to the simulation results recorded in the SAIF file.

TLPM Usage

Power compiler is a primary tool in the TLPM power characterization process. TLPM utilizes the main functionality for power consumption evaluation. A SAIF file is cleared from any activity, while signal activity is added separately to evaluate the power consumption of each signal. Then, the signals are added incrementally to study the correlation among them.

4.4   TLPM Implementation and Simulation

After completion of power characterization process on different design components, correlation matrix is built. TLPM methodology uses the correlation matrix in implementation phase in order to log power parameters into power models. Implementation phase is initiated with the mapping mechanism of the correlation matrix of RTL to TLM. Power models are built to track the activity in the design elements using read/write operations and attach the adequate energy information to them. TLPM implementation is summarized in Fig. 4.10.

TLPM implementation is based on building of the adequate power models for registers using the correlation matrix. This is achieved by mapping of the RTL signals to TLM design registers. Power evaluation is done upon simulation of TLM which encloses power models. Power models track the activity of the design and assign proper power numbers for each register operation. At the end of simulation, total power consumption is reported.

Fig. 4.10 TLPM
implementation process

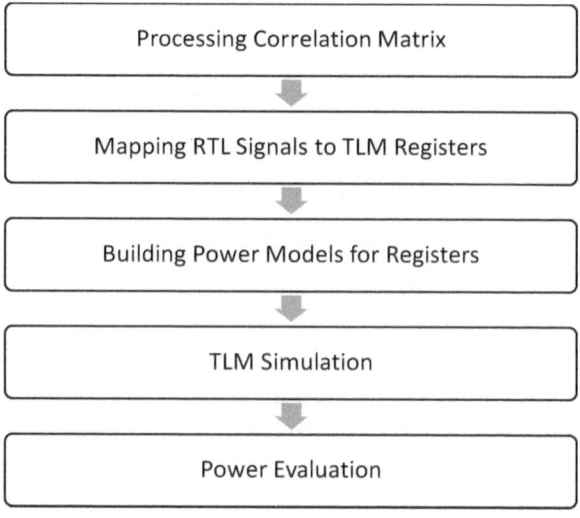

The design model in TLPM methodology is PVT. PVT model contains both functionality and timing of a design unit. For timing, approximately timed modeling at high level of abstraction of the design components is adopted to achieve fast simulation with high precision.

4.4.1 TLPM Implementation and Simulation Flow

The design model in TLPM methodology is PVT. PVT model describes both functionality and timing for the design units. For timing, approximately timed modeling at high level of abstraction is adopted to achieve fast simulation with high precision.

Implementation phase starts with mapping the RTL signals to the TLM registers. The correlation matrix is applied to the relevant TLM registers. The power model is created for every read/write transaction to track the activity of the design registers. Thus, the exerted energy for every register is attached to the registers. Total energy is evaluated at end of simulation.

The power models of the design registers are implemented using different TLM ports. The initiator and debug ports are used in the implementation. On calling the registers, the power model is executed and energy is evaluated. Some RTL signals do not have corresponding TLM registers. Separate function is needed to model this power consumption.

4.4.1.1 Signal-to-Register Mapping

Power models are implemented inside TLM registers. Every source of power dissipation in RTL needs to be mapped to TLM to evaluate power correctly. For the signal-to-register mapping scheme, a debugging tool is needed. Visualizer debugging tool is used for this purpose. The signal tracking technique is adopted for the mapping scheme. The technique is performed by tracking the drivers and the receivers of signals to the design ports.

Visualizer causality features are used for signal tracking technique. The Logic Cone view is used to trace signal drivers and receivers. The drivers and receivers are traced to the ports or the interface ports themselves. Time Cone view is used to track signal values and changes in behavior across time.

After checking the mapping scheme for the design signals in RTL and corresponding TLM registers, they are categorized as follows:

1. Signals exist in RTL and do exist in TLM description. These are represented of TLM as registers and ports. For these signals, power evaluation is done on read/write operations of the registers.
2. Signals exist in RTL, but are implicitly implemented in TLM description. Clock signals are examples of those. There are signals/nets for clocks in RTL, but there is no explicit corresponding registers or ports in TLM description.

3. Signals exist in RTL, with no implementation at all in TLM description. Signals
   of power management components or internally generated control signals are
   examples of those. Those signals are not used during the normal operation of
   TLM. But, TLPM methodology considers their activities.

4.4.1.2   Power Evaluation for TLM Registers

TLPM methodology evaluates the power in signals through two types of ports in
TLM: initiator ports and debug ports.

Initiator Ports

They represent the real behavior of communication for the model. They are the
interface between the design unit and the whole system design of SoC. The communication between TLM model of a single component and SoC is done through
callback functions. Callback functions are implemented for every read/write operation of each register.

Power models are built for the registers. Power equations are added to these
callback functions to evaluate the power upon every read/write operation. The power
equation is tracking the activity of the registers. Upon a call of a register read/write
operation, the toggle count of the activity increases. Hence, energy exerted due to
this register increases.

After annotating the RTL contributors to the model, the power model is built using
the following steps:

1. Track the activity of TLM registers.
2. Calculate the exerted energy by every operation.
3. Accumulate the energy result over the previously calculated values. As simulation
   time advances, the model gathers all exerted energy across the design.

A power model in TLM includes two sections:

1. Evaluation for changes in values registers.
2. Power evaluation for the corresponding changes.

A snippet of the first section of power model is added in Fig. 4.11. It shows the
first section of callback function for a register. This function represents the write
functionality of a load register. When the register is called, a new value which is
supplied as an argument is assigned to the register. The model evaluates the following:

1. Number of calls for this function has been done till this time step.
2. Number of bins/bits has changed from the last call. This is performed by XORing
   the binary value from the current call and the previous one.

A snippet of the second section of power model is added in Fig. 4.12. It shows the
calculation of power exerted on a register. It holds the values of exerted power per
change in the register. The changes could be either:

```
// This is write callback for TimerlLoad register.
// The newValue has been already assigned to the TimerlLoad register.
// newValue is supplied as an argument; reloaded immediately to start operation.

void TIMER_SP804_pv::cb_write_TimerlLoad(unsigned int newValue) {

    // Section: Evaluation for changes in bins
    static unsigned int TimerlLoad_call_count = 0;    // To count how many calls had been done
                                                      // for register since operation start.
    static unsigned int TimerlLoad_keep = 1;          // To keep the register previous value.
                                                      // To check how many bins have changed
    static unsigned int TimerlLoad_bins = 0;          // Accumlation for how many bins have changed
    unsigned int TimerlLoad_new = newValue;           // Assigning the new value argument to local var
    unsigned int result;                              // To evaluate the number of changes in bins

    TimerlLoad_call_count++;                          // Incrementing the calls number for this callback function

    result = TimerlLoad_keep ^ TimerlLoad_new;        // Evaluation of number of bins changes on new call

    if (result > 0) {                                 // if there is a change in load value than last call
                                                      // then we need to evaluate it
        TimerlLoad_bins = TimerlLoad_bins + popcount(result);  // Extracting number of bins changes and add it back
        TimerlLoad_keep = TimerlLoad_new;             // Setting the keep value to teh new call value
    }
```

Fig. 4.11 Power model—tracking of register changes

```
// Section: Evaluation of power numbers
unsigned int power_load, power_total, power_call_count;
                                  // power_load: signals exerted power on changing bins
                                  // power_call_count: signals exerted power on calls
                                  // power_total: total exerted power

unsigned int power_load_bin = 66;       // unit power for single bin change
unsigned int power_call_count_bin = 3200;  // unit power for calls

power_load = power_load_bin * Timer1Load_bins;
power_call_count = power_call_count_bin * Timer1Load_call_count;
                                  // Mutiplying the power info and corresponding count
power_total = power_load + power_call_count ;  // Addition and saving the information
power_save(power_total);

  m_timers[0]->load_value(newValue);
}
```

Fig. 4.12 Power model—evaluation of power

1. Exerted power due to the call of the register.
2. Exerted power due to the change of bits.

Both items are based on the corresponding signals if they contribute due to just the call itself or the changes in bits are the fundamental contributor.

After evaluation of the local power of this register, the information is saved in a global variable for data collection and the total power is evaluated.

Debug Ports

They represent the backdoor access to the TLM design. TLPM methodology uses both initiator and debug ports in power estimation. The user can access information of all registers using the debug port without interfering the blocking transport calls on initiator port.

Initiator ports track activity of one register at a time. Multiple registers need to be tracked at the same instant to estimate power accurately. Thus, the debug ports are needed along with the initiator ports to track multiple registers. However, they need explicit calls in the stimulus function besides the initiator ports.

An example for a debug port is added in Fig. 4.13. It shows the call of debug port of an interface. The information is accessed through a function call. The function accepts two arguments:

1. Memory address of the register: This address is based on the TLM implementation and the used peripheral.
2. Variable: This would hold the read information for the mentioned memory address.

Further, same processing is performed here similar to processing on initiator port.

4.4.1.3 Power Evaluation for Non-TLM Signals

For signals that exist in RTL with no implementation or with implicit implementation in TLM, an additional function is implemented to evaluate power of those signals according to their contribution factors. These factors are extracted from the correlation matrix which is generated from power characterization step.

An example for this additional stand-alone function is presented in Fig. 4.14. The function includes an example for signals with implicit TLM implementation (e.g., clocks). The power is evaluated by energy exerted for single change, the current simulation time value and the clock value itself. On the other hand, for RTL signals that have no implementation in TLM (e.g., control signals) are also considered in this function. The power is also evaluated by the energy exerted for single change and the current simulation time value. The evaluated power values are then accumulated to have the total value for the function.

```
// Calls for Debug Interface Ports
// AMBA APB interface is used here

AMBA_APB_read_dbg (0x00,rData);

AMBA_APB_read_dbg (0x04,rData);
AMBA_APB_read_dbg (0x08,rData);
AMBA_APB_read_dbg (0x18,rData);
AMBA_APB_read_dbg (0x20,rData);
AMBA_APB_read_dbg (0x24,rData);
AMBA_APB_read_dbg (0x28,rData);
AMBA_APB_read_dbg (0x38,rData);
```

```
// This function call takes two arguments
// First argument:  Memory address for register
// Second argument: Variable name that holds the info
```

Fig. 4.13 Debug port example

```
unsigned int power_calculate_standalone()
{
    // Current Time
    double time_now = sc_time_stamp().to_default_time_units();

    // Clocks Power Values
    unsigned int power_bin_pclk = 404;
    unsigned int power_sampling_pclk = (float(power_bin_pclk) * 1000000.0 * 2.0 ) / float(apb_clock);
    unsigned int power_bin_timclk = 230;
    unsigned int power_sampling_timclk = (float(power_bin_timclk) * 1000000.0 * 2.0 ) / float(tim_clock);

    // Calculate STATIC (RTL only) power (non-functionalty power, not related to data values)
    unsigned int power_bin_static_prescale = 98;
    unsigned int power_static_prescale = power_bin_static_prescale * time_now ;

    unsigned int power = power_sampling_pclk + power_sampling_timclk + power_static_prescale;

    return power;
}
```

Fig. 4.14 Stand-alone function for non-TLM signals

4.4.1.4   TLPM Simulation

Upon finishing the mentioned procedure for power models implementation, the design is ready for TLPM simulation. While running simulation, different design operations are triggered for execution for their communication across components. The activity of the registers is tracked in power models. Energy exerted per operation is calculated. At the end of simulation, all energy components are accumulated and dynamic power is concluded.

4.4.2   Power Model Example

Considering the correlation matrix example mentioned before. The RTL and the corresponding TLM description are described in Fig. 4.15. The design has two input ports "input_1" and "input_2". They are affecting the internal signals in the design. According to the RTL description, the first datapath affected by "input_1" has signal "a". On the other hand, signals "b" and "c" on the second datapath are being affected by "input_2".

Fig. 4.15 Power model

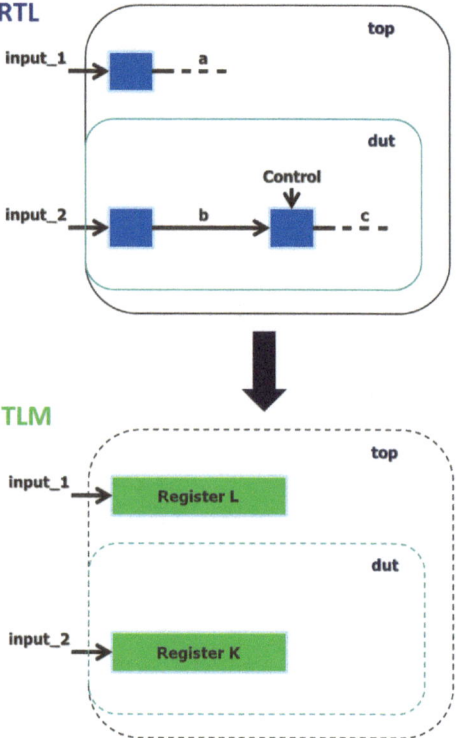

Thus, for the TLM description, the first datapath is associated with register "L", while the second one is associated with register "K". As mentioned before, on performing power characterization process on this design, signal "a" has no dependency on any other signal for the value of exerted energy. This value is the same through all configuration changes.

While for signal "b" located under "top/dut", it is subject to configuration changes, because an extra load "c" is added to the datapath. The signal depends on a "Control" signal value.

An example is illustrated in Algorithm 4.2. The evaluation of the power model for register "L" is based on signal "a", while for register "K" is based on signals "b" and "c". According to the example, the energy per toggle for a register is a function of corresponding signals in RTL description. The toggles for certain register is evaluated by comparing/xoring the previous value to the newer one. Then, the exerted energy is evaluated by multiplication of this calculated toggle count and the energy per toggle. Finally, this information is logged to total energy function which handles all registers power model functions.

Algorithm 4.2 Power Model Example

Registers: 'L' $\Rightarrow$ depends on signal 'a' only while 'K' $\Rightarrow$ depends on signal 'b' and 'c'.

function **reg_K_callback** (new_value)
    Energy$_{toggle}$ (K) = Energy$_{toggle}$ (b) + Energy$_{toggle}$ (c)
    TGL = count_toggles_function (new_value)
    Energy (K) = Σ (TGL) * Energy$_{toggle}$
    total_energy_func (Energy (K))
end function

function **count_toggles_function** (new_value)
    count_toggle = new_value   XOR   previous_value
    return (count_toggle)
end function

4.4.3 EDA Tools for TLPM Implementation Phase

Different tools are exploited in TLPM implementation process. Those tools are presented briefly with respect to the common flow and specific needed usage.

4.4.3.1 Visualizer ®

Overview

Visualizer [6] is a high-performance debugger tool by Mentor (Siemens). It is integrated with QuestaSim tool for digital design and verification. It has the ability for

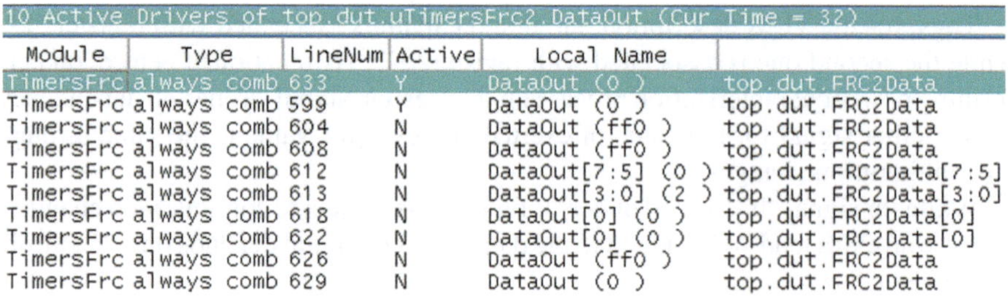

```
10 Active Drivers of top.dut.uTimersFrc2.DataOut (Cur Time = 32)
┌──────────────┬──────────────┬─────────┬────────┬─────────────────┬──────────────────────┐
│ Module       │ Type         │ LineNum │ Active │ Local Name      │                      │
├──────────────┼──────────────┼─────────┼────────┼─────────────────┼──────────────────────┤
│ TimersFrc    │ always comb  │ 633     │ Y      │ DataOut (0 )    │ top.dut.FRC2Data     │
│ TimersFrc    │ always comb  │ 599     │ Y      │ DataOut (0 )    │ top.dut.FRC2Data     │
│ TimersFrc    │ always comb  │ 604     │ N      │ DataOut (ff0 )  │ top.dut.FRC2Data     │
│ TimersFrc    │ always comb  │ 608     │ N      │ DataOut (ff0 )  │ top.dut.FRC2Data     │
│ TimersFrc    │ always comb  │ 612     │ N      │ DataOut[7:5] (0 )│ top.dut.FRC2Data[7:5]│
│ TimersFrc    │ always comb  │ 613     │ N      │ DataOut[3:0] (2 )│ top.dut.FRC2Data[3:0]│
│ TimersFrc    │ always comb  │ 618     │ N      │ DataOut[0] (0 ) │ top.dut.FRC2Data[0]  │
│ TimersFrc    │ always comb  │ 622     │ N      │ DataOut[0] (0 ) │ top.dut.FRC2Data[0]  │
│ TimersFrc    │ always comb  │ 626     │ N      │ DataOut (ff0 )  │ top.dut.FRC2Data     │
│ TimersFrc    │ always comb  │ 629     │ N      │ DataOut (0 )    │ top.dut.FRC2Data     │
└──────────────┴──────────────┴─────────┴────────┴─────────────────┴──────────────────────┘
```

Fig. 4.16 Visualizer drivers/receivers window

waveforms analysis, connectivity, and causality checking on different HDL such as SystemVerilog, VHDL, and SystemC. Visualizer supports different features such as UVM, TLM, RTL, and GL.

Standard Flow

The tool has two modes of operation, live-simulation mode, and post-simulation mode. The steps are performed as follows:

1. Visualizer design hierarchy file is created during QuestaSim optimization step.
2. Simulation logs the design signal values using proper options.
3. Visualizer is launched using the design hierarchy file and signals logged data.

TLPM Usage

For TLPM implementation process, power models are built for registers. They are supplied with energy values for power evaluation. TLM registers need to be mapped to the relevant power contributors in RTL description, where power values are logged for RTL signals in the correlation matrix. To map signals to registers, signal tracking technique is adopted. Visualizer debugger tool facilitates the tracking technique through different causality features.

The purposes of tracking are as follows:

1. Following the drivers and the receivers of all signals to the effective design ports.
2. Logging the behavior of signal drivers over simulation time.
3. Tracking the TLM registers to the design ports.

Checking the drivers of a signal is shown in Fig. 4.16. It shows the drivers for a signal called "DataOut". There are ten drivers for that signal and only two of them are active drivers at specific time "$t(n)$".

Signal tracking is done visually to the design ports using Logic Cone window, as shown in Fig. 4.17. This window shows the dataflow of RTL signals. The user can trace back for the drivers and forward for the receivers. This is similar for schematic representation for RTL.

This figure shows drivers and receivers for a signal called "Load". The drivers are tracked to "PCLK", "LoadEn", and "PWDATA". "PWDATA" is tracked to its

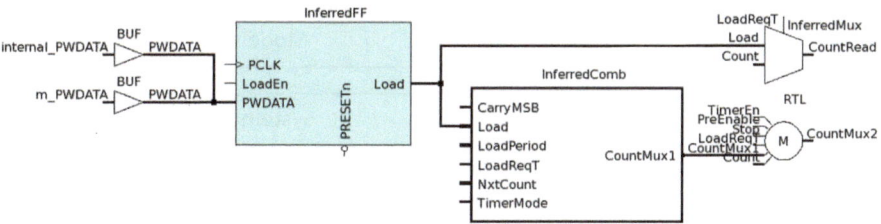

Fig. 4.17 Visualizer Logic Cone window

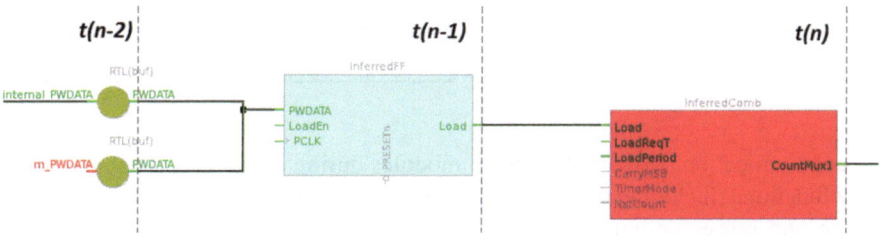

Fig. 4.18 Visualizer Time Cone window

early driver ports with UVM environment interface. Also, receivers for "Load" are "CountRead" and "CountMux1".

However, the Logic Cone window gives full representation for schematic, but user may need to check the relation between signals versus their drivers and receivers over time. In order to check the effect of certain signal at certain time step, Time Cone window is used as shown in Fig. 4.18.

In this figure, Time Cone is shown for signal "CountMux1" at time "$t(n)$". The changes in the drivers are tracked through time "$t(n − 1)$", then back to "$t(n − 2)$" for the earliest change. After checking all signals using this approach, every signal is attached to a register. The power value is added for power model of that register.

4.4.3.2 Vista ®

Overview

Vista [7] is a TLM-based solution from Mentor (Siemens). It is a native Electronic System Level (ESL) solution for design modeling, assembly, virtual prototyping, and verification. Vista provides system architects with full view of the design to effectively create TLM models for validation, power, and performance analysis. Hardware designers are able to validate critical changes early in the design phase before RTL implementation.

Vista has TLM2.0 generic model library for models for various processors such as ARM, PPC, and MIPS, in addition to peripherals models such as cache, snoop, and interconnect models. The model builder in Vista is used to create models for

Fig. 4.19 Vista flow

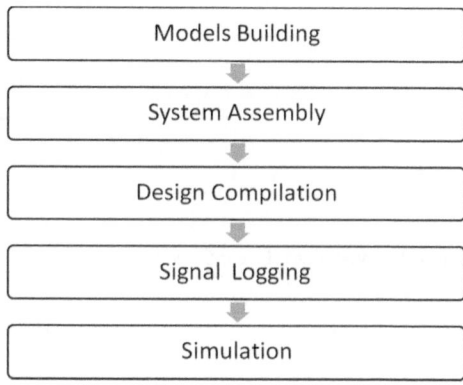

the communication between the design modules during simulation. The models are fully functional including timing and power policies. Those policies model complex timing and power definition for different design modules in intuitive way. Vista builds models onto two layers: programmers' view (PV) and timing (T) layers. PV contains the functional behavior of the component. T layer captures the timing and power behavior. The two layers represent clear distinction between functionality and timing. For timing behavior, the user chooses between loosely timed, approximately timed, or cycle-accurate models with the same functionality for every T model.

During TLM simulation using Vista, a database is created for traced transactions. Users are able to analyze the transactions throughput, power consumption, and delays using Graphical User Interface (GUI). Vista has the ability to create a virtual prototype file for stand-alone execution. This file includes simulation executables, required shared libraries, and configuration settings for general parameters of a specific core.

Standard Flow

Vista has intuitive user interface for the whole flow, starting from the creation of the needed models till the analysis of the simulation output.

The flow is illustrated in Fig. 4.19 and it is performed in the following steps:

1. The library of the design units is created using ModelBuilder.
2. CPU, memory, and bus models are instantiated from Vista generic library.

   (a) Ports, PV functionality, and T policies are defined through the GUI.
   (b) The following set of files is generated for every component:
       i. PV: for the models functionality.
       ii. T: for timing and power policies.
       iii. Model: has the model infrastructure.

3. Top level is created by building system architecture through component assembly.
4. Design project is compiled.
5. Rebuilding/compiling the project with addition of needed models and top level.
6. Signals to be traced are logged.

7. Simulation is launched.

   (a) User is able to change T model between approximately timed (AT) or loosely timed (LT).

TLPM Usage

Vista represents the core of TLPM methodology. The tool provides the TLM flow of TLPM power estimation methodology. The usage is listed as follows:

1. Vista facilitates building all design components:

   (a) The user chooses the proper design components according to the system architecture and specifications.
   (b) Every component is configured to contain the design ports and needed power and time policies to match design specifications.

2. TLM design analysis and verification:

   (a) Vista provides utilities for design debugging. It gives transaction analysis with respect to throughput and latency.

3. Vista facilitates the developing of the instrumentation functions:

   (a) Power estimation in TLM requires upgrading the generated models to suite TLPM methodology.
   (b) Vista enables a proper models upgrade with syntax and semantics verification for the introduced implemented functionality.

4. Power estimation of TLPM models:

   (a) Power models are introduced in registers callback functions.
   (b) During simulation, the registers are triggered and the energy is evaluated for every transaction.
   (c) At the end of the simulation, energy exerted by all operations is accumulated and power is evaluated.

4.5 Summary

In this chapter, the TLPM methodology for power estimation is discussed in detail. The motivation of TLPM implementation is presented at first. TLPM is an efficient methodology for power estimation at high level of abstraction.

The flow of TLPM is presented in detail. TLPM consists of two stages: power characterization and TLPM implementation. Characterization represents the extraction of power parameters from existing design. Implementation represents the addition of power model to TLM and the execution of this model to have reliable numbers for dynamic power consumption. CAD tools for every stage are explained.

Power characterization process is explained in detail. It uses "Power Analysis flow" for the extraction of power parameters and the cross relation between the signal power. The power parameter is defined as the exerted energy by the RTL signal for a single toggle count. A correlation matrix is built after deducing the power parameters of RTL signals and their interdependency.

TLPM uses the correlation matrix in the implementation phase. RTL signals are mapped to the corresponding TLM registers. Correlation matrix information is supplied to those TLM registers. Power models are built in callback functions of registers. They are accessed using two types of ports: initiator and debug ports. Extra function is implemented for signals with no actual implementation in the TLM description.

Upon simulation of the modified TLM design, power models are tracking the activity of operations. The energy exerted by every design component is calculated. Total dynamic power value is estimated at the end of the simulation.

References

1. Perl.org. The Perl Programming Language: Perl Documentation. http://www.perl.org.
2. Mentor Graphics, Inc. QuestaSim User Guide. https://www.mentor.com.
3. IEEE. 1801-2015 - IEEE Standard for Design and Verification of Low-Power, Energy-Aware Electronic Systems. http://www.ieee.org.
4. Accellera Systems Initiative. Universal Verification Methodology. http://www.accellera.org.
5. Sengupta, S., Saurabh, K., Allen, P.E. (2004) A process, voltage, and temperature compensated CMOS constant current reference. In *Proceedings of the 2004 International Symposium on Circuits and Systems*, Vol. 1, (pp. I–I).
6. Mentor Graphics, Inc. Visualizer Debug Environment. https://www.mentor.com.
7. Mentor Graphics, Inc. Vista User Guide. https://www.mentor.com.

Chapter 5
Experimental Results

5.1 Introduction

TLPM qualification has been performed on different designs. Power characterization phase creates the correlation matrix which encloses the design power parameters of signals. Correlation matrix is used for implementation of power models in the TLM description. Power models are added in the callback functions of registers. When a register is accessed during simulation, power models evaluate the change and calculate the exerted energy. In this book, TLPM is applied on Timer SP804 [1] and the ZYNQ-7000 SoC [2], in order to prove the efficiency of the methodology. In this chapter, experimental results are added for power estimation on TLPM. Accuracy and speed are compared for TLM and RTL results.

5.2 Design Environment

UMC technology kit is used in the synthesis/characterization process [3]. Library specifications are listed in Table 5.1.

5.3 Timer SP804 Experiment

5.3.1 Design Information

The used design under test is ARM Dual-Timer Module (SP804) [4]. The design has two similar Free Running Counters (FRC). The design is connected to the core using Advanced Peripheral Bus (APB). A block diagram for the timer is shown in Fig. 5.1. A detailed design schematic is generated by Visualizer tool as shown in Fig. 5.2.

© Springer Nature Switzerland AG 2020
A. B. Darwish et al., *Transaction-Level Power Modeling*,
https://doi.org/10.1007/978-3-030-24827-7_5

Table 5.1 UMC technology library

Feature	Value
Process	Typical
Temperature	25 °C
Voltage	1 V
Model	fsd0a_a_generic_core_ttlv25c
Technology	90 nm

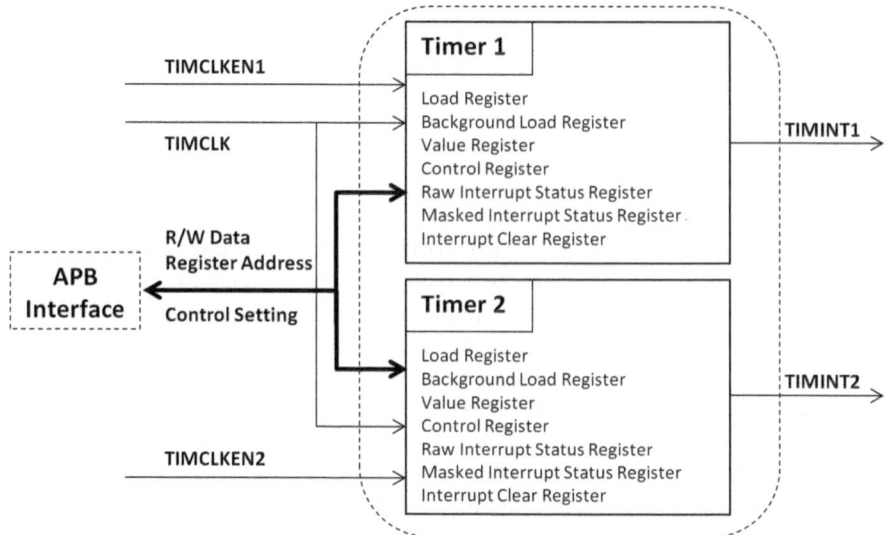

Fig. 5.1 ARM Dual-Timer Module (SP804)

The design has a common clock for both timers with separate clock enables for flexible control. When the counter reaches zero, an interrupt signal is generated. The interrupt can be cleared by internal interrupt clear register. An external interrupt signal can also be applied after being masked-off.

5.3.2 Design Configuration

The dual timers can be programmed with the following configurations:

1. Programmable timer size:
   It can operate at 16 bit or 32 bit.
2. Programmable counting mode:

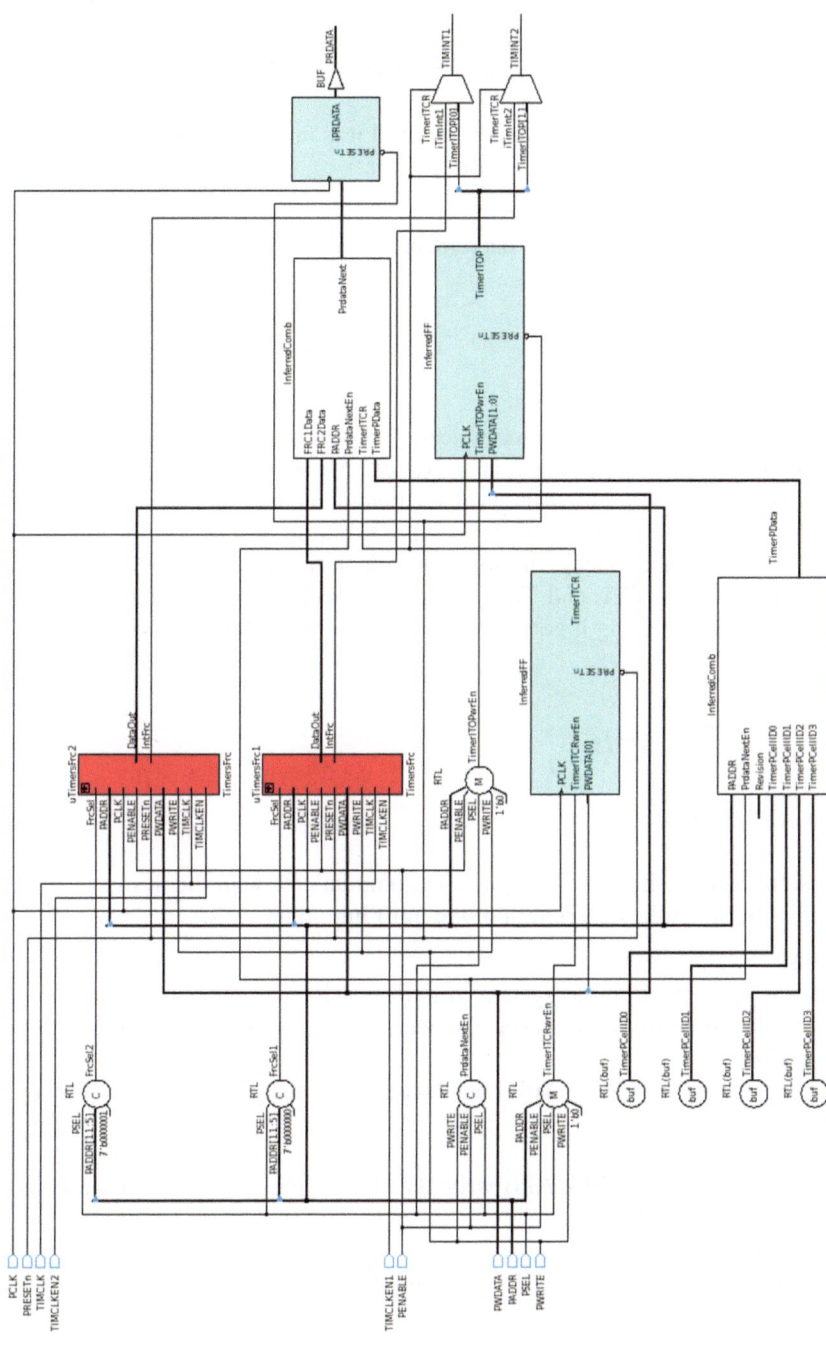

Fig. 5.2 Schematic for ARM Dual-Timer Module (SP804)

    (a) Free running mode: Timer counts in this mode continuously. Counter decrements till it reaches zero, then wraps to its configured maximum value.
    (b) Periodic mode: Timer counts in this mode continuously as well. But the counter here wraps to the value written to load register, when counter reaches zero.
    (c) One shot: This mode counts once until it is reprogrammed again. Counter starts with value written to load register.

3. Programmable pre-scale divider:
   This feature controls the count decrement rate of the Timer: 1, 16, or 256.

5.3.3   Design Interface

Design interface facilitates the communication schemes between memories, processors, and peripherals. In this work, an Advanced Microcontroller Bus Architecture (AMBA) [5] interface protocol is used for communication between core and the timer. AMBA has a set of protocols to serve different design specifications. The protocols vary in performance, the supported clock frequency, and bus size. In this work, APB is used in RTL and TLM design as interconnected with the core. APB has low bandwidth making it good candidate for the timer design.

5.3.4   Design Signals

AMBA APB Signals

The design interface has several signals to communicate with the design as shown in Fig. 5.3. Below is the description of the most important ones:

1. PCLK:
   It is the main clock in the system. It is responsible for all sampling across the design. It is used by APB interface to access the registers in the design.
2. PRDATA:
   It is unidirectional AMBA APB read data bus. It is responsible for passing information to the core.
3. PWDTA:
   It is unidirectional AMBA APB write data bus. It passes the data from the core to the timer.
4. PENABLE:
   It is the enable signal for AMBA APB interface.

Fig. 5.3 AMBA APB
signals

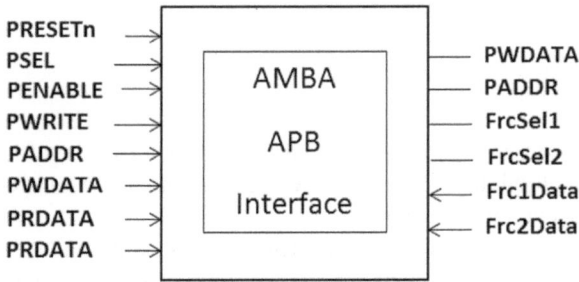

Non-AMBA APB Signals

1. TIMCLK:
   It is the clock of operation for the dual-timer counters. TIMCLK should comply
   with the following rules:

   (a) TIMCLK should be synchronized with PCLK.
   (b) TIMCLK should be equal to or multiple of PCLK, but cannot be faster than
       PCLK.

2. TIMCLKEN1, TIMCLKEN2:
   They are the clock enable for each timer.
3. TIMINT1, TIMINT2:
   Interrupt output is generated for each timer when the count reaches zero.

5.3.5   Registers of the Design

Dual-timer design has a set of identical registers for each timer. The critical ones are
listed as follows:

1. TimerXLoad
   It is a read/write register. When it is reloaded, the counter restarts from this new
   value immediately. The behavior is used in periodic and one-shot modes.
2. TimerXValue
   A read-only register containing the current timer value.
3. TimerXControl
   It is a read/write register, responsible for several design configuration attributes.
   It controls timer enable, mode, size and pre-scale.
4. TimerXBGLoad
   It has only one difference with TimerXLoad. When it is reloaded, the counter
   does not restart immediately.
5. TimerXIntClr
   When this write-only register is updated, the interrupt is cleared.

5.3.6 TLPM Results

5.3.6.1 Test Plan

Experimental results are performed to verify the efficiency of the proposed TLPM methodology. The power values on both TLM and RTL are compared to check the accuracy of the approach.

Also, simulation execution time is determined for every scenario to elaborate the effectiveness of the methodology. ARM Dual-Timer SP804 has different sets of inputs, controls signals, and clocks. Multiple stimulus scenarios are used to cover all registers of the timer. They operate the design at different clock frequencies and with different control settings to achieve sufficient coverage. The test plan is listed in Table 5.2. The assumed scenarios are listed in Table 5.3.

5.3.6.2 Power Results

A comparison for power estimation is performed for TLPM and RTL. The results for both TLPM and RTL are shown in Fig. 5.4.

Table 5.2 Test plan for ARM Timer SP804

Feature	Value
PCLK (ns)	2
	4
	8
TIMCLK (ns)	2
	4
	8
	16
Mode	FRC
	Periodic
	One-shot
Size (bit)	16
	32
Pre-scale	1
	16
	256
Number of load/BG-load write instructions	No
	Few
	Excessive

Table 5.3 Test scenarios

Scenario	PCLK	TIMCLK	Mode	Size	Pre-scale	No. of load/BG-load instructions
1	2	2	FRC	32	1	No
2	2	2	Periodic	32	1	No
3	2	2	FRC	16	1	Excessive
4	2	2	Periodic	16	1	Excessive
5	2	4	FRC	32	16	Few
6	2	4	Periodic	16	16	Few
7	2	4	FRC	32	1	Few
8	2	4	Periodic	16	1	Excessive
9	2	4	FRC	32	16	Excessive
10	2	4	FRC	16	16	Excessive
11	2	8	FRC	16	1	Excessive
12	2	8	Periodic	16	1	Excessive
13	2	2	Periodic	16	256	Excessive
14	2	8	FRC	32	1	Few
15	2	8	Periodic	32	1	Few
16	2	2	FRC	16	1	Few
17	2	16	Periodic	32	1	No
18	4	4	Periodic	32	1	No
19	4	4	FRC	16	1	Excessive
20	4	4	Periodic	16	1	Excessive
21	4	8	FRC	32	16	Few
22	4	8	Periodic	16	16	Few
23	4	8	FRC	32	1	Few
24	4	8	Periodic	16	1	Excessive
25	4	16	FRC	32	16	Excessive
26	4	16	Periodic	16	16	Excessive
27	4	16	FRC	32	1	Excessive
28	4	16	Periodic	16	1	Excessive
29	8	8	FRC	32	1	Excessive
30	8	8	FRC	16	1	Few
31	8	8	FRC	16	1	Few
32	8	8	Periodic	16	1	Few
33	8	16	Periodic	16	1	No
34	8	16	FRC	32	1	No
35	8	16	Periodic	32	1	Few
36	8	16	FRC	32	1	Few

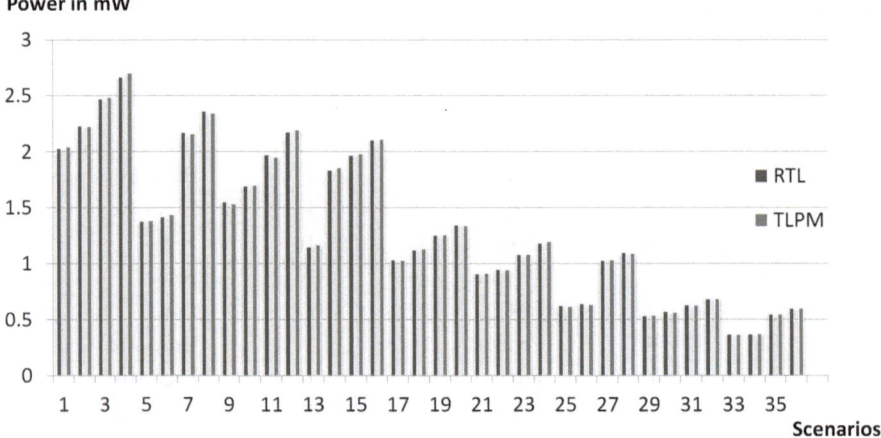

Fig. 5.4 Dynamic power estimation for different operating conditions

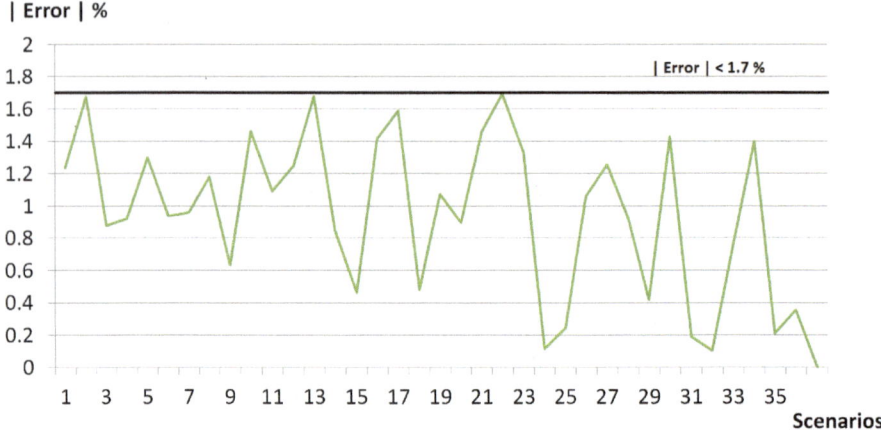

Fig. 5.5 Dynamic power estimation absolute error between TLPM and RTL

The power estimation absolute error between TLPM and RTL is maximum 1.7%, as illustrated in Fig. 5.5.

5.3.6.3 Execution Time

There are two factors in the evaluation of execution time of TLPM methodology. The first factor is the time consumed during power characterization stage for design blocks. The second factor is the simulation time in TLM models including the power models.

Time (sec)

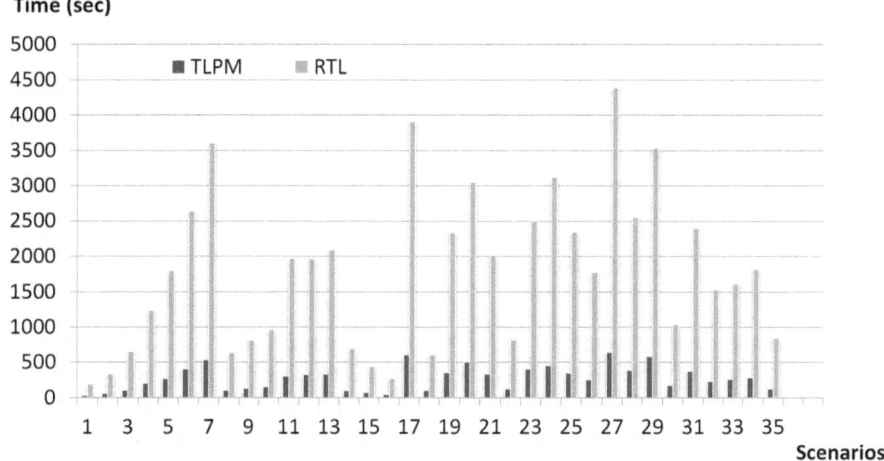

Fig. 5.6 Simulation time for different operating conditions

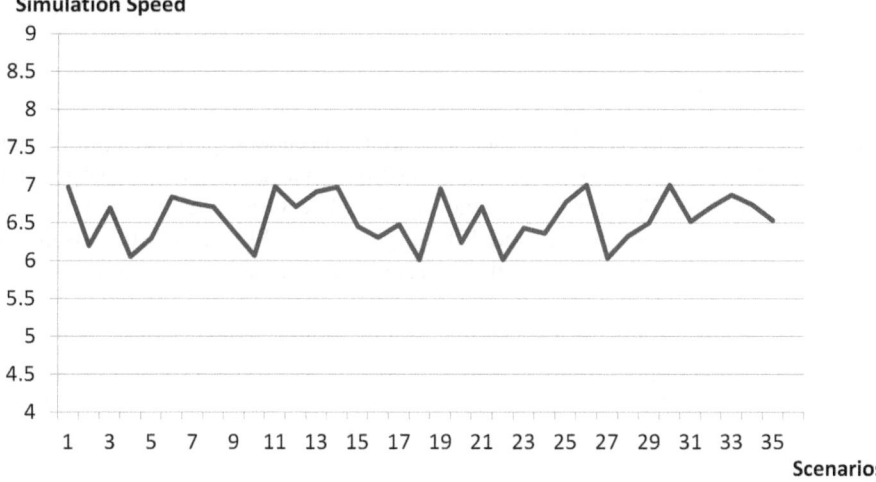

Fig. 5.7 Simulation time comparison between TLPM and RTL

The process power characterization on SP804 Timer is a semiautomated process. The process takes 30 min to complete the generation of the correlation matrix.

On the other hand, simulation time of each executed scenario on TLM is presented in Fig. 5.6.

TLPM methodology achieves upto $7\times$ speedup in simulation time as compared to RTL as illustrated in Fig. 5.7.

The significant advantage of using TLPM could not be demonstrated on small block design. The speedup in simulation for small devices could reach single order of magnitude.

5.4 ZYNQ-7000 SoC Experiment

TLPM added more value is illustrated by a large design. For large SoCs, simulation with RTL could be not feasible. However, TLPM could achieve orders of magnitude speedup on large designs.

5.4.1 Design Information

The used design under test is ZYNQ-7000 [6] shown in Fig. 5.8. Xilinx All Programmable SoC (AP SoC) is the base for ZYNQ-7000 family. This family consists of two systems: Processing System (PS) and Programmable Logic (PL). Dual-core ARM Cortex-A9 MPCore and some peripherals exist in the PS. The FPGA design is implemented in PL. In the following subsections, some blocks are used as examples for applying TLPM on ZYNQ. The details of these blocks (GPIO, SPI, I2C, and UART as examples) of the SoC along with the modeling aspects are discussed.

5.4.1.1 GPIO

The General-Purpose I/O (GPIO) is user programmable general-purpose input/output controller. It is mainly used to implement functions that require simple output and input software-controlled programmable signals. The GPIO peripheral provides access to 64 inputs from the PL and 128 outputs to the PL through the extended multiplexed I/O (EMIO) interface. It also provides observation and control of up

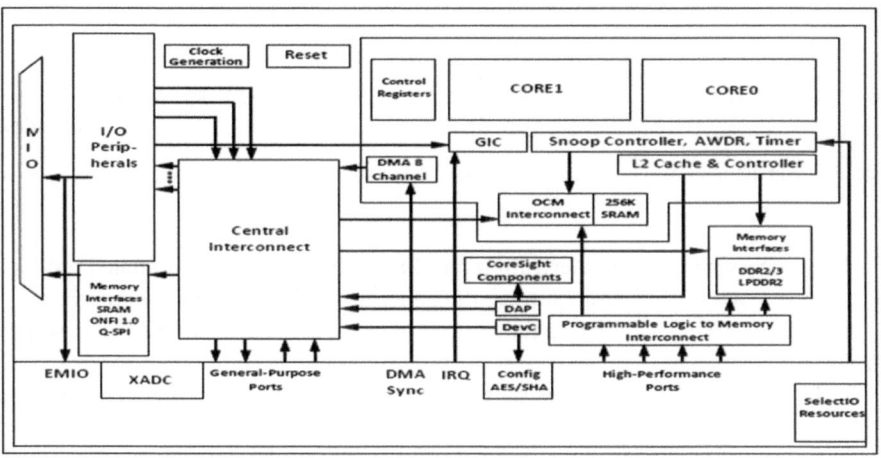

Fig. 5.8 ZYNQ-7000 platform

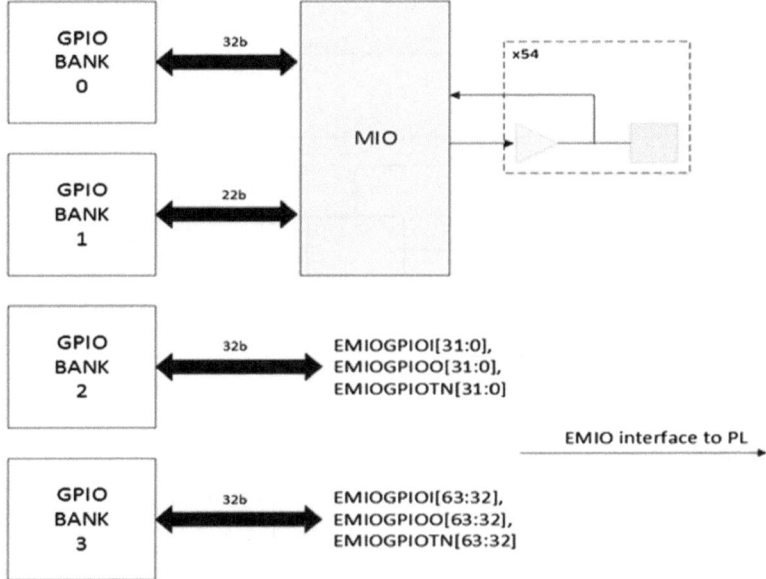

Fig. 5.9 GPIO block diagram

to 54 device pins via the Multiplexed I/O (MIO) module. The GPIO is organized into four banks of registers that group related interface signals. Each GPIO can be dynamically and independently programmed as input, output, or interrupt sensing. Software can read all GPIO values within a bank using a single load instruction, or write data to one or more GPIOs using a single store instruction. The block diagram of the GPIO is illustrated in Fig. 5.9.

The GPIO channel starting from the software-configured registers till the pins which describe the operation of Bank0 and Bank1 is illustrated in Fig. 5.10.

The power dissipated in GPIO during reading/writing basically occurs from/to the programming registers. The power dissipation of the read/write operation of the registers is common in all devices and is calculated in the same way. In the callback function of each register, the power dissipation due to read/write is calculated.

For the register operations, there are two types of correlation given below that are considered:

1. Full correlation: The bits of each register have the same power contribution. They have a correlation factor of ONE. The energy per unit toggle per bit is added to the write and read callback function of each register. The energy exerted by APB signals (the programming port/protocol of the registers of the device) is also added and taken into consideration.
2. Partial correlation: Each register has partial correlation with APB signals. Conditional correlation is applied. The main APB signals that affect the power of read/write operations are PSEL, PRESETn, PWRTIE, and PREAD.

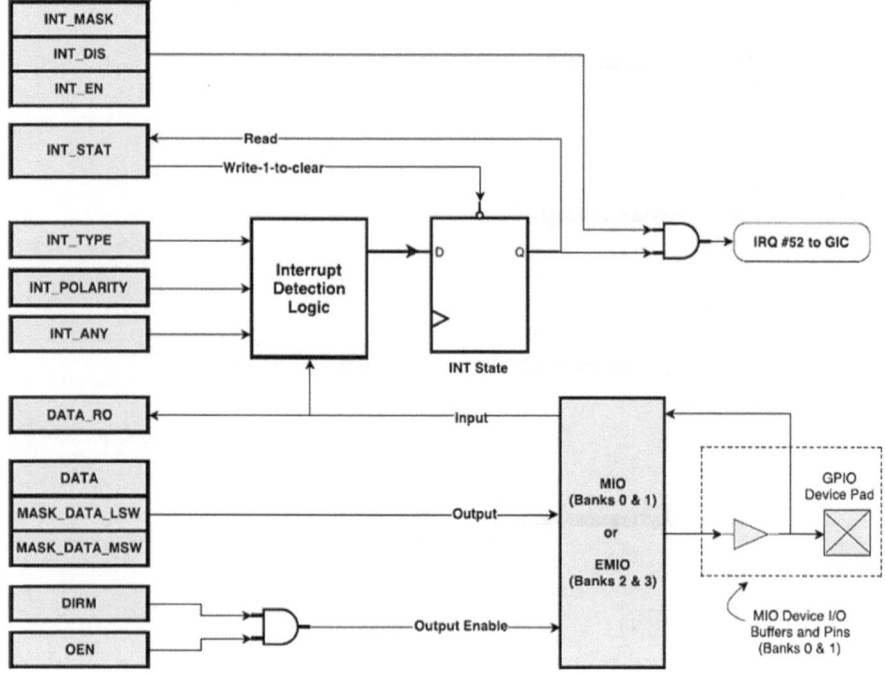

Fig. 5.10 GPIO channel

5.4.1.2   SPI

Serial communication with many peripherals such as temperature sensors, memories, SD card, analog converters, displays, pressure sensors, and real-time clocks is performed using the Serial Peripheral Interface (SPI) that supports full duplex communication between master and slave. SPI device supports up to three slaves. It also drives the SPI reference clock that simulates the baud-rate division of the system reference clock. The system reference clock can be divided by (4, 8, 16, 32, 64, 128, and 256). Delay between the transmitted bytes can also be added.

SPI can operate in two modes: master mode and slave mode. In master mode, data flow from APB bus to Transmitter First In First Out (TXFIFO) then to Master Output Slave Input (MOSI) data bus, while in slave mode data move from APB to TXFIFO to Master Input Slave Output (MISO). At the same time, in master mode, data move from slave to MISO then Receiver First In First Out (RXFIFO) to APB bus. In slave mode, data move from slave to MOSI to RXFIFO to APB bus. The power dissipation of SPI module is mainly divided to

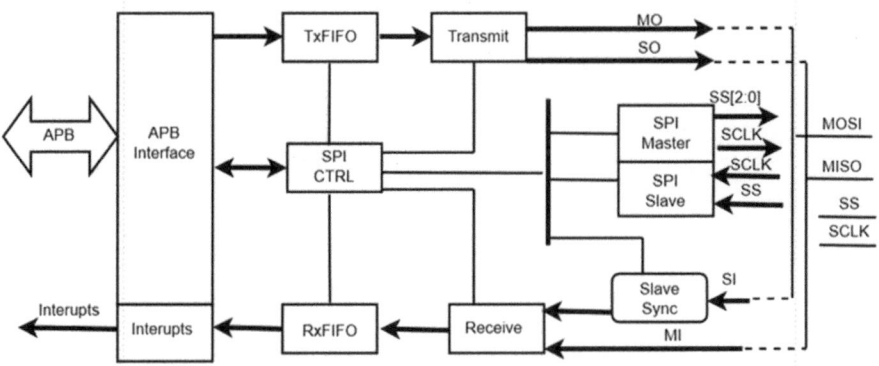

Fig. 5.11 SPI block diagram

1. Power dissipation due to read/write operations on registers as described in GPIO.
2. Clock power: The clock has ZERO correlation with other signals. That means its power is independent of any other signal.
3. Clock dividers power: The power of clock dividers is calculated each time a byte is sent or received. The number of toggles per clock enable cycle is calculated. The energy exerted by the clock dividers during sending/receiving a byte is calculated from the following equation:

$$EB = NC*NT*ET, \qquad (5.1)$$

where EB: energy exerted by clock dividers per byte, NC: number of baud-rate cycles per byte, NT: number of toggles per baud-rate cycle, and ET: the energy per unit toggle in counters of the clock dividers.

The calculated number (EB) is added every time a byte is sent or received. EB is accumulated in the write callback function of the receiving port (DATA IN). The start of transmission can be configured to be manual or automatic. In the manual mode, the transmission starts when the software writes in a specific bit in the configuration register. In automatic mode, whenever a byte is written to the TXFIFO, transmission starts. The block diagram of the SPI is illustrated in Fig. 5.11.

5.4.1.3 I2C

Inter Integrated Circuit (I2C) device can be configured to run in different modes (master transmitter, master receiver, slave transmitter, and slave receiver). It also supports slave monitoring mode which can monitor the slave until it is ready. Moreover, clock stretching is supported. The address can be configured for 7 bits or 10

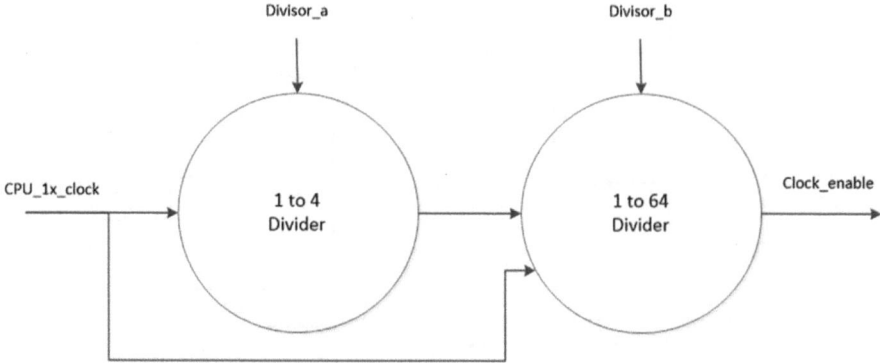

Fig. 5.12 Baud-rate generator of I2C

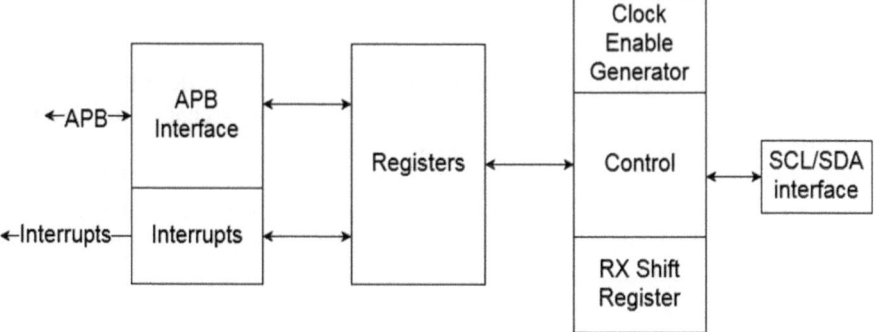

Fig. 5.13 I2C block diagram

bits. Also, it has the ability to configure the baud rate. The baud rate is generated by dividing the CPU clock by two configurable values as shown in Fig. 5.12.

Increasing the dividing value decreases the transmitting and receiving rate of I2C. In I2C, the transmitter must read an ACK from the receiver to go on with the transaction, otherwise the transmitter stops. Thus, master and slave are used during testing to complete the appropriate handshaking [7]. The block diagram of the I2C module is shown in Fig. 5.13.

The power dissipation components of I2C module is similar to that of SPI. There is independent power dissipation for the clock signal. There is also power for read/write from/to registers. For the power of the clock dividers, EB is calculated as explained in the SPI. In master mode, whenever an address is written to the address register, the transaction starts. In the write callback function of the address register, there is a for loop that which performs the transaction for one iteration per byte. EB is accumulated to the total energy at each iteration of this loop. In slave mode, the transactions go through I2C slave ports. EB is accumulated in the callback functions of those ports in the read callback in slave receiver mode and write callback in slave transmitter mode.

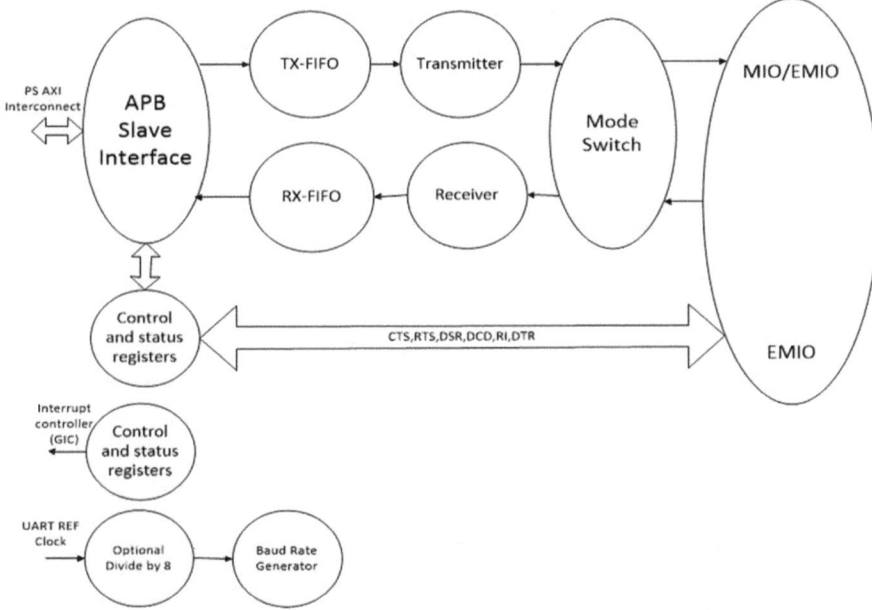

Fig. 5.14 UART block diagram

5.4.1.4 UART

Universal Asynchronous Receiver Transmitter (UART) is used for different kinds of serial communication. The design of the UART is implemented according to the block diagram in Fig. 5.14.

It supports four modes with different baud rates. The modes are normal, automatic echo, local loopback, and remote loopback. The block encloses three divisions for the clock and varying number of bytes to be sent or received along the different modes. The key factors that affect the simulation time and power dissipation are the number of bytes and the divisors of the clock to be used for transmitting and receiving. The reference clock can be divided by three values as shown in Fig. 5.15.

Increasing the dividing value decreases the transmitting and receiving rate of UART. The resulted value is the baud rate of UART. For the power of the clock dividers of UART module, energy per byte (EB) is calculated as mentioned before. EB is accumulated at the write callback function for the receiving port. For transmission, EB is accumulated at the write callback function for the transmission FIFO (TXFIFO). The platform blocks are connected together through APB bus. It is used to select the required peripheral and handles the transactions between these blocks and the ABP bus.

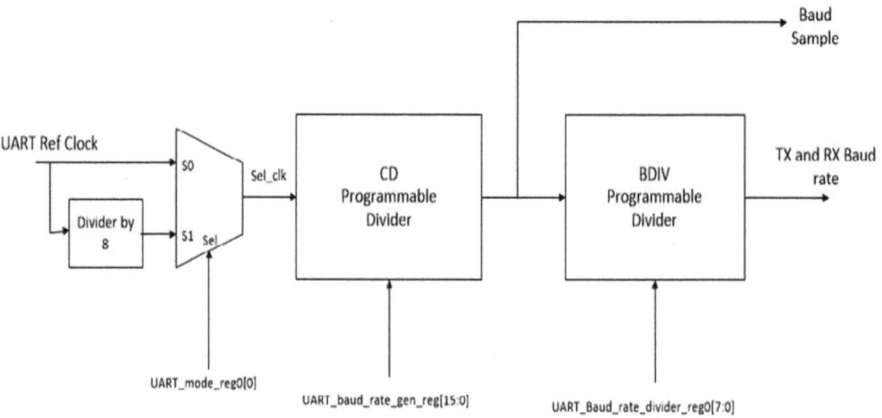

Fig. 5.15 Baud-rate generator of UART

5.4.2   TLPM Results

5.4.2.1   Test Plan

The exercised scenarios for effective modules:

GPIO:

GPIO has one main operation scenario. Configuration of MIO pins is done by storing specific values in GPIO registers. To write data to output pins, two options are used as follows:

1. Option 1: update the GPIO pins using the gpio. DATA_0 register.
2. Option 2: The MASK DATA x MSW/LSW registers, which mask the most/lowest significant half of the bank, are used to update one or more GPIO pins.

While in order to read from input pins, two approaches are performed as given below:

1. gpio.DATA_RO_x register of each bank is used and read by APB master.
2. Interrupt logic is used on input pins.

SPI:

Experiments are developed along changing the clock division. Power is calculated in RTL and TLM to demonstrate how much the power of the TLM model matches that of the RTL model.

I2C:

To test the power model of I2C, two modules of I2C are used. One is configured as master and the other is configured as slave. Different number of bytes in addition to different clock divisions are used to estimate power.

UART:

A normal mode for transmission is tested with different configurations through changing number of bytes and the clock division.

The Experimental Scenarios for the Full System:

Scenario A:

The SPI master is configured to transmit 256 bytes which are received by the slave. The APB clock is divided by 32. The I2C is set to master receiver and slave transmitter mode. The slave is configured to transmit 256 bytes. In order to achieve the required baud rate, the reference clock is divided by two values 1 and 10. The reference clock is finally divided by 10. The UART is configured to normal mode. It is configured to transmit 8 bytes. The reference clock is divided by eight and two configurable dividers. In this scenario, the dividers are configured to 255. The clock is finally divided by 8 * 255 * 255 = 520,200.

Scenario B:

For SPI, clock is divided by 64. The master of I2C is configured to transmit 512 bytes. Clock is divided by two values 2 and 20. The reference clock is finally divided by 2 * 20 = 40. The UART is configured to transmit 16 bytes. The clock dividers are configured to divide the clock by 255 and 511. The clock is finally divided by 8 * 255 * 511 = 260,610.

Scenario C:

The master of I2C is configured to transmit 1024 bytes. The clock is divided by two values 3 and 31. The reference clock is finally divided by 3 * 31 = 93. The UART is configured to transmit 32 bytes. The clock dividers are configured to divide the clock by 255 and 1023. The clock is finally divided by 8 * 255 * 1023 = 2,086,920.

Scenario D:

For SPI, the clock is divided by 256. The master of I2C is configured to transmit 2048 bytes. The clock is divided by two values 3 and 63. The reference clock is finally divided by 3 * 63 = 189. The UART is configured to transmit 64 bytes. The clock dividers are configured to divide the clock by 255 and 4095. The clock is finally divided by 8 * 255 * 4095 = 8,353,800.

5.4.2.2   Power Results

GPIO:

Power values for RTL and TLM are shown in Table 5.4. It is shown that error in estimating power dissipation of GPIO is less than 2%.

Table 5.4 Dynamic power estimation for different operation conditions of GPIO

Power dissipation (uW)		Error (%)
RTL	TLM	
258.02	253.23	1.85

Table 5.5 Dynamic power estimation for different operation conditions of SPI

Clock division	Power dissipation (mW)		Error (%)
	RTL	TLM	
4	1.145	1.129	1.33
8	1.024	1.021	0.312
16	0.940	0.940	−0.039
32	0.890	0.892	−0.231
64	0.856	0.863	−0.803
128	0.839	0.845	−0.736
256	0.830	0.837	−0.896

Table 5.6 Dynamic power estimation for different operation conditions of I2C

Clock division	Number of bytes	Power dissipation (uW)		Error (%)
		RTL	TLM	
144	10	177.02	176.00	0.5
3	50	179.29	176.00	1.4
1	30	182.50	178.00	2.2

SPI:

Power values for RTL and TLM are shown in Table 5.5. For different operating scenarios, absolute error in power estimation of SPI is less than 1.5%.

I2C:

Table 5.6 shows the power obtained from both RTL and TLM for different configurations. Different number of bytes in addition to different clock divisions are used to estimate power. For different operating scenarios, error in power estimation for I2C is less than 3%.

UART:

Testing is performed through different configurations by changing number of bytes and the clock division as shown in Table 5.7. For different operating scenarios, absolute error in power estimation is less than 1%. High accuracy in power estimation for different TLM models when compared with RTL is achieved using TLPM making it efficient for high-level power estimation.

Table 5.7 Dynamic power estimation for different operation conditions of UART

Clock division	Number of bytes	Power dissipation (uW)		Error (%)
		RTL	TLM	
8	15	142.43	143.00	−0.39
15	15	143.93	144.00	−0.33
255	5	143.18	143.00	0.57

Table 5.8 Dynamic power estimation for different operation conditions of ZYNQ-7000

Scenario	Power dissipation (mW)		Error (%)
	RTL	TLM	
A	1.286	1.260	2
B	1.271	1.259	1
C	1.265	1.258	0.5
D	–	1.257	–
Average	1.274	1.259	1.1

The Experimental Scenarios for Full System:

Total power is calculated for these scenarios in RTL and TLM as shown in Table 5.7. The error in power estimation of TLM does not exceed 2%. The power of Scenario D could not be estimated on RTL as the VCD file which is used in power estimation exceeds 1TB which could not be stored on normal PC. RTL simulation time of this scenario exceeds 80h. Using TLPM, power estimation of Scenario IV becomes feasible.

5.4.2.3 Execution Time

Simulation time is also calculated as shown in Table 5.9. The simulation time in RTL increases rapidly as the clock division increases. In TLM, the simulation time is constant because the clock is abstracted with a variable in the transaction level. Scenario D becomes feasible with simulation time of 66 s as illustrated in Tables 5.8 and 5.9. TLM provides accurate power estimation with faster simulation. TLPM is efficient for power profiling. A speedup in simulation time using TLPM of up to 628 is achieved. On average, the speedup in different scenarios is almost 290. RTL fails to simulate complicated scenarios. Unlike RTL, TLM not only achieves more than two orders of magnitude speed up in simulation but also allows heavy SW to run and the power to be estimated accurately.

Table 5.9 Simulation time for different operating scenarios of ZYNQ-7000

Scenario	Simulation time (s)		RTL/TLM ratio
	RTL	TLM	
A	1616	66	24.4
B	14,400	66	216.1
C	41,437	66	627.8
D	–	66	–
Average	19,151	66	289.499

5.5 Summary

TLPM methodology is proposed for dynamic power estimation using TLM. TLPM is applied on ARM Dual-Timer Module (SP804) design and on ZYNQ-7000 SoC. The methodology speeds up the simulation of different scenarios by up to 628 times while keeping the error in power estimation to be on average less than 3%. More than two orders of magnitude speedup in simulation to determine power estimation for large SoC are achieved with TLPM.

References

1. Darwish, A., El-Moursy, M., & Dessouky, M. (2016) Transaction level power modeling (TLPM) methodology. In *2016 17th International Workshop on Microprocessor and SOC Test and Verification (MTV)* (pp. 61–64), IEEE.
2. Baher, A., El-Zeiny, A. N., Aly, A., Khalil, A., Hassan, A., Saeed, A., et al. (2018). Dynamic power estimation using transaction level modeling. *Microelectronics Journal, 81*, 107–116. November.
3. United Microelectronics Corporation. http://www.umc.com.
4. ARM Ltd. ARM Dual-Timer Module (SP804) Technical Reference. http://www.arm.com.
5. ARM Ltd. ARM AMBA Advanced Microcontroller Bus Architecture. http://www.arm.com.
6. Xilinx. Zynq-7000 All Programmable SoC Technical Reference Manual. https://www.xilinx.com.
7. I2C Bus Specifications. https://www.i2c-bus.org.

Chapter 6
Conclusions and Future Work

6.1 Conclusions

In this book, different SoC design challenges and their resolution are discussed. The advance in processing technologies has derived the complexity of electronic design. Evaluating power consumption at early phase of product life cycle is important to decrease the number of expensive design iterations. Different CAD tools and modeling techniques have been developed to continuously support building new SoCs. TLPM is a new methodology for dynamic power estimation using Transaction-Level Modeling. TLPM is used to overcome the design challenges of long design iterations in electronic design as an efficient solution for dynamic power estimation for large SoC.

TLPM methodology exploits the existing tools for RTL simulation, design synthesis, and SystemC prototyping to provide fast and accurate power estimation. TLPM is feasible and can be deployed easily in the industry. TLPM consists of two stages: Power characterization and TLPM implementation. Characterization represents the extraction of power parameters from the existing design. It uses "Power Analysis flow" for extraction of power parameters and cross correlation of design signals. The power parameter is the exerted energy by the RTL signal for single toggle count. Correlation matrix is built after deducing the power parameters of RTL signals and their dependency on each other.

TLPM uses correlation matrix in the implementation phase. Implementation represents the addition of power model to TLM and the execution of this model to have reliable numbers of dynamic power consumption. Correlation matrix information is supplied to the TLM registers after mapping RTL signals to the corresponding registers. Power models are built in callback function of the registers. Upon simulation of TLM design, the power models are tracking the activity of the operations. The energy exerted by every design component is calculated. Total dynamic power value is estimated at the end of the simulation.

The methodology is applied first on the block level. Simulation of the full SoC with real software scenarios is then performed. Implementation of commercial tools

© Springer Nature Switzerland AG 2020   107
A. B. Darwish et al., *Transaction-Level Power Modeling*,
https://doi.org/10.1007/978-3-030-24827-7_6

and flows is performed. The efficiency of TLPM is demonstrated. The methodology speeds up the simulation of different scenarios by up to 628 times while keeping the error in power estimation to be on average less than 3%. More than two orders of magnitude speedup in simulation to determine power estimation for large SoC are achieved with TLPM. Product life cycle could be enhanced using TLPM to estimate the power consumption at early design phases.

6.2   Future Work

Following this book, TLPM methodology can be deployed and updated as follows:

1. Apply TLPM on various designs.
2. Enhancement for TLPM methodology for better performance results. The plan is to enhance evaluation technique in power models for better speed results. The current implementation requires frequent calls in the test bench in order to have accurate evaluation of registers' activity. The new proposal is to upgrade the power models to have less frequent register calls. This is performed through creation of smart power models that interpret the intermediate register activity without the need of frequent calls.
3. Include static power estimation in TLPM methodology.
4. Full automation for characterization process.

Index

A

Abstraction levels, 1, 10, 17, 25, 47, 49, 53
Accuracy, 4, 11, 12, 18, 19, 25, 31, 33, 49,
   50, 53, 54, 68, 87, 92, 104
Approximately timed, 14, 18–20, 25, 27–29,
   31, 32, 36, 37, 49, 50, 54, 73, 84, 85
Approximately timed coding style, 18
Automation, 108

B

Base classes, 40
Base protocol, 19, 28, 35–37
Blocking transport interfaces, 20, 26
Blocking versus non-blocking transport in-
   terfaces, 20
Bus bridge example, 23

C

CAD, 53, 85, 107
Circuits, 1, 2, 10, 65
Coding style, 14, 17–20, 24, 26, 27, 29, 31–
   36
Coding styles and use cases, 20
Coding styles in TLM, 17
Combined TLM-2.0 interfaces, 34
Component level, 3, 48
Contribution, 2, 3, 47, 49, 63, 65, 77, 97
Correlation, 61–63, 65, 66, 68, 72, 73, 80,
   82, 86, 87, 95, 97, 99, 107
Correlation matrix example, 66
Cycle-accurate modeling, 20

D

Datasheet, 51

Datasheet-based characterization, 51
Debugger, 81, 82
Debug transport interface, 14, 22–24, 26,
   33–35
Design abstraction levels, 10
Design compiler, 69
Design configuration, 88, 91
Design environment, 87
Design information, 11, 87
Design interface, 90
Design modeling, 9, 50, 83
Design signals, 44, 71, 73, 82, 90, 107
Design synthesis, 1, 9, 43, 45, 61, 69, 107
Differences between coding styles, 19
Direct Memory Interface (DMI), 24, 32
DMI interface, 23

E

EDA, 53, 66, 81
EDA tools for power characterization, 66
EDA tools for TLPM implementation phase,
   81
Efficiency, 11, 87, 92, 108
Energy, 2, 3, 50–53, 61–63, 65, 66, 68, 72–
   74, 77, 80–82, 85–87, 97, 99–101,
   107
Enhancement, 108
Execution time, 68, 92, 94, 105

F

Field Programmable Gate Array (FPGA),
   53, 96

© Springer Nature Switzerland AG 2020
A. B. Darwish et al., *Transaction-Level Power Modeling*,
https://doi.org/10.1007/978-3-030-24827-7

G
Gate level, 1, 10, 43, 50
Gate-level netlist characterization, 50
Gates, 1, 2, 9, 10, 43, 50, 52–54
General-Purpose I/O (GPIO), 96, 97, 102
Generation of signal switching activity, 53

H
Hardware characterization, 50
Hardware Description Language (HDL), 1,
       9, 12, 44, 68, 82

I
Integrated Circuits (IC), 1, 99
Interfaces, 10, 14, 17, 19–24, 26–28, 31–36,
       38, 39, 41–44, 73, 74, 77, 84, 90, 96,
       97
Inter Integrated Circuit (I2C), 96, 99, 100,
       102–104

L
Levels of abstraction in modeling, 49
Levels of abstraction in TLM, 25
Look-up tables, 52
Look-up tables' characterization, 52
Loosely timed coding style and temporal de-
       coupling, 17

M
Message chart—blocking transport, 26
Message chart—temporal decoupling, 27
Message chart—time quantum, 27
Message sequence—early completion, 29,
       31
Message sequence—timing annotation, 29
Message sequence—using backward path,
       28
Message sequence—using return path, 28
Motivation, 85

N
Netlist, 2, 43, 47, 50, 53, 54, 62, 63, 69, 71
Non-blocking transport interface, 14, 19, 20,
       24, 26–28, 36

O
Object/component proxies, 41, 42
Optimization, 2, 68–70, 82

P
Performance, 2, 9, 10, 52, 54, 68, 81, 90, 108
Performance overhead, 52
Permitted phase transitions, 36
Phase sequences, 35
Power, 1–3, 9, 44, 47–54, 61–63, 65, 66, 68–
       74, 77, 80–87, 92, 94, 97, 98, 100–
       108
Power analysis, 53, 54, 63, 86, 107
Power analysis flow, 53, 86, 107
Power calculation, 53
Power characterization, 2–4, 47, 50, 53, 54,
       61–63, 65, 72, 77, 95, 107
Power estimation, 3, 47, 48, 61, 62, 85, 87,
       94, 104–108
Power estimation using TLPM, 62
Power evaluation, 2, 49–53, 72–74, 77
Power evaluation for non-TLM signals, 77
Power evaluation for TLM registers, 74
Power model, 3, 4, 48–52, 54, 61–63, 72–74,
       80–83, 85–87, 94, 102, 107, 108
Power model example, 80, 81
Power modeling, 1–3, 47, 61
Power modeling at different levels, 47
Power modeling in TLM, 3
Predefined classes, 42
Product design cycle, 1, 9, 10, 49
Purpose, 12, 24, 26, 27, 45, 68, 73, 82, 96

Q
QuestaSim, 68

R
Registers of the design, 91
Register Transfer Level (RTL), 1, 2, 9–12,
       26, 43, 48, 53, 54, 61–63, 65, 66, 68,
       69, 71–74, 77, 80, 82, 86, 92, 96, 102,
       104, 105, 107

S
Scope, 3, 12
Serial Peripheral Interface (SPI), 96, 98–
       100, 103, 104
Signal-to-register mapping, 73
Simulation, 1, 2, 4, 10–12, 14, 17–19, 23,
       24, 26, 27, 32, 34, 39, 41–45, 48–52,
       54, 61–63, 66, 68, 71–74, 77, 80, 82,
       84–87, 92, 94–96, 101, 105–108
Sockets combination, 23

Switching Activity Interchange Format
    (SAIF), 9, 44, 45, 53, 63, 65, 66, 68,
    71, 72
Switching between approximately timed and
    loosely timed, 19
Synthesis, 1, 9, 43, 45, 53, 61, 69, 107
System level, 2, 12, 21, 44, 47, 48, 83
System-on-Chip (SoC), 1–3, 10–12, 14, 44,
    48, 54, 63, 74, 87, 96, 106, 107
SystemC, 1, 9, 12, 14, 18–20, 29, 31, 34, 44,
    51, 52, 61, 68, 82, 107
SystemC example, 12

T
Temporal decoupling example, 18
Timer, 17, 18, 87, 88, 90–92, 95
Timer SP804 experiment, 87
Timing annotation, 26, 28, 29, 31–33
TLM classes, 14
TLM components, 21
TLM design abstraction, 11
TLM interfaces, 23, 24
TLM ports, 73
TLM-2.0 global quantum, 34
TLM 2.0 interfaces, 26
TLM-2.0 phases, 35
TLM-2.0 sockets, 34
TLPM flow, 61
TLPM implementation and simulation, 73
TLPM methodology, 3, 9, 47, 61, 63, 68, 69,
    72–74, 77, 85, 92, 94, 95, 106–108
TLPM power characterization flow, 63
TLPM simulation, 80

Transaction-Level Modeling (TLM), 1–3, 9–
    11, 14, 17, 20, 21, 24–26, 35, 44, 48–
    50, 52, 54, 61–63, 68, 72–74, 77, 80–
    87, 90, 92, 95, 102–107
Transaction-Level   Power   Modeling
    (TLPM), 2–4, 9, 44, 47, 49, 61–63,
    66, 68, 69, 72–74, 77, 80–82, 85–87,
    92, 94–96, 104–108
Transaction lifetime example, 21
Transistor size, 1, 2
Transition sequence flowchart example, 36
Transport interface and DMI comparison, 33
Transport interfaces, 20, 26, 33

U
Universal Asynchronous Receiver Transmit-
    ter (UART), 96, 101, 103, 104
Universal   Verification   Methodology
    (UVM), 9, 39–43, 45, 68, 83
Untimed coding style, 17
UVM factory, 41

V
Verification, 2, 9, 25, 39, 40, 45, 51, 52, 68,
    81, 85
Vista, 83
Visualizer, 81

Z
ZYNQ, 87, 96
ZYNQ-7000 SoC experiment, 96